12/19

D0472711

LEARN TO PROGRAM
WITH APP INVENTOR

LEARN TO PROGRAM WITH APP INVENTOR

A VISUAL INTRODUCTION TO BUILDING APPS

BY LYRA BLIZZARD LOGAN

no starch press

San Francisco

Printed in USA

First printing

23 22 21 20 19   1 2 3 4 5 6 7 8 9

ISBN-10: 1-59327-968-X
ISBN-13: 978-1-59327-968-4

Publisher: William Pollock
Production Editor: Meg Sneeringer
Cover Illustration: Josh Ellingson
Developmental Editor: Annie Choi
Technical Reviewer: Abigail Joseph
Copyeditor: Rachel Monaghan
Compositor: Meg Sneeringer
Proofreader: James Fraleigh
Indexer: Beth Nauman-Montana

For information on distribution, translations, or bulk sales, please contact No Starch Press, Inc. directly:
No Starch Press, Inc.
245 8th Street, San Francisco, CA 94103
phone: 1.415.863.9900; info@nostarch.com
www.nostarch.com

Library of Congress Cataloging-in-Publication Data

Names: Logan, Lyra, author.
Title: Learn to program with App inventor : a visual introduction to building apps / Lyra Logan.
Description: San Francisco : No Starch Press, Inc., [2019]
Identifiers: LCCN 2019020522 (print) | LCCN 2019021958 (ebook) | ISBN
    9781593279691 (epub) | ISBN 1593279698 (epub) | ISBN 9781593279684 (print)
    | ISBN 159327968X (print)
Subjects: LCSH: Application software--Development. | Android (Electronic
    resource) | Mobile computing. | Mobile apps. | Smartphones--Programming.
Classification: LCC QA76.76.A65 (ebook) | LCC QA76.76.A65 L64 2019 (print) |
    DDC 005.3--dc23
LC record available at https://lccn.loc.gov/2019020522

To Mom, who instilled the passion to learn and to teach

ABOUT THE AUTHOR

Lyra Blizzard Logan is a certified Perl, JavaScript, and Database Design Specialist and Web Development Professional who develops web applications using PHP and MySQL. She is an MIT Master Trainer in Educational Mobile Computing (App Inventor) and holds graduate certificates in Instructional Technology: Web Design and in Online Teaching and Learning. She earned her BA from Fisk University and JD from Harvard. Currently, Logan is Executive Vice President and General Counsel of the Florida Education Fund (FEF), a nonprofit that creates and implements educational programs for underrepresented groups. In addition, Logan directs FEF's pre-college programs, which include after-school and summer coding camps for elementary, middle, and high school students in Florida.

ABOUT THE TECHNICAL REVIEWER

Abigail Joseph holds a PhD in Computer Science with an emphasis on data visualization. For 15 years she has been teaching technology, computer programming, and design thinking to middle school students and K–12 educators in the San Francisco Bay Area, with a focus on web design, Processing Graphical Library, JavaScript, Scratch, App Inventor, and more. Beyond teaching, making, and the arts, her current passion is *edstoria.com*, which uncovers ways to help prevent teachers from vanishing from the educational landscape. She tweets at @drabigailjoseph.

BRIEF CONTENTS

CONTENTS IN DETAIL

ACKNOWLEDGMENTS

Can a corporate attorney write a coding book in plain enough English for kids to understand? Only with the help of the team at No Starch Press. Thanks to publisher Bill Pollock and editors Annie Choi, Meg Sneeringer, and Rachel Monaghan for helping to make these words less and less complex draft after draft. Thanks to Dr. Abigail Joseph for technically reviewing the manuscript and helping write its accompanying lesson plans. Thanks also to artist Josh Ellingson for a fantastic cover.

I could not have written this book without Stephen Kamau, who planted the notion years ago that I could learn PHP; Lawrence Morehouse, who allowed me to start and expand FEF's out-of-school coding programs, which have become my laboratory for exposing underserved youth to computer programming; and my mother, the late Lora Ann Archey, who steered me away from the teaching profession, but inspired me through her life's work in education to somehow find a way to teach.

I would not have finished this book without the enthusiastic encouragement of my late brother Clarence "Chip" Blizzard III, and niece Karen Isabella Blizzard, who proudly asked about my progress almost every day, and my husband Dr. Willie Logan, who has been my sounding board, mentor, and most ardent supporter for close to thirty years.

INTRODUCTION

Technology influences all aspects of our lives, and computing jobs and opportunities abound. It's more important than ever for everyone to understand how computers work. It's especially important that kids, including those from underrepresented groups, know that they have the ability to write software, study computer science, or become computer programmers if they so choose.

In this book, you'll learn how to make useful and exciting *mobile software*—apps for phones and tablets—with App Inventor, while exploring key programming concepts along the way. We'll create apps that send text messages, recognize speech, access a phone's list of contacts, sense the phone's location, operate the camera and camcorder, solve math problems, play video and sound, animate graphics, turn text into speech, and even respond to touch and let users draw on the screen!

Many young people around the world are using App Inventor to create apps like these to positively impact their communities. Kids in California developed an app that allows users to record and report graffiti and organize clean-up events. North Dakota middle school students created an app to encourage more recycling in their community. Middle school girls in Texas designed an app that uses the Global Positioning System (GPS) to guide visually impaired students around their school. Teens in Nigeria built an app that uses location sensing to help traffic cops catch offenders. Young women in Europe designed a crowdsourcing app to help residents find safe drinking water. A middle school boy in India developed an app to help parents locate their child's school bus and check if their child is on the bus. The possibilities for creative problem solving are endless!

As an app developer, you too will have the power to solve real-world problems and help your community using your smartphone. You'll use your critical thinking and problem-solving skills to develop apps that tell the computer exactly what you want it to do.

WHAT IS APP INVENTOR?

App Inventor is a free online visual programming environment originally developed by Google and now maintained by the Massachusetts Institute of Technology (MIT). It empowers people to create software for their phones and tablets rather than simply use those devices.

In traditional programming environments, you need to type the actual code to develop software in a language like PHP or Java. But App Inventor lets developers create technology relatively quickly by dragging and dropping code blocks onto the screen rather than typing out code. This approach makes it easy for anyone to create and publish a simple app for a smartphone or tablet in under an hour. You can also use App Inventor to program more complex apps with high-tech mobile computing features in significantly less time than you can with traditional text-based programming environments. With App Inventor, you'll be able to develop software in no time—and with fewer errors too!

Best of all, App Inventor blocks incorporate the same programming logic you would use if you were writing code in text-based environments. This means that the programming concepts you learn here will help you as you start to code with traditional languages.

WHO SHOULD READ THIS BOOK?

This book is for curious coders age 11 and up and any parents or educators who want to introduce kids to programming. The book, along with its supplementary materials and lesson plans, is designed to demystify coding, show you how to build cool apps, and expose you to computational thinking skills and fundamental programming concepts.

This book is also for anyone who wants to build fun, feature-rich apps for mobile devices without getting bogged down by the tedious nuances of a text-based language. Not only will you become familiar with App Inventor and learn how to build the apps in this book, you'll also be able to create your own powerful apps to share with friends or upload for sale.

WHAT'S IN THIS BOOK?

In this book you'll learn how to use App Inventor's drag-and-drop interface to create a fun app in each chapter. Along the way, you'll learn computational thinking skills like decomposing problems into manageable parts and fundamental computer programming concepts like handling events, storing data in variables and lists, using control structures to direct program flow, creating procedures to perform tasks, refactoring code, and debugging.

The "On Your Own" exercises at the end of each chapter test your understanding of the concepts and prepare you to create your own apps. These problems also reinforce your knowledge of programming principles that can ease your transition to traditional text-based coding.

Here's what you'll find in each chapter:

Chapter 1: Building Apps with App Inventor   You'll create the "Hi, World!" app, which lets you text a message via speech recognition to a telephone number from your contact list, and then package the app to send to your friends to install.

Chapter 2: App Inventor and Event-Driven Programming   Here you'll use event handlers to create an app you can use to record, watch, and re-record temporary video clips of yourself practicing a speech or song.

Chapter 3: Fruit Loot: Creating a Simple Animated Game   Program animations and use variables to build the "Fruit Loot" game, where you earn points for catching randomly dropping fruit.

Chapter 4: Multiplication Station: Making Decisions with Code   Use conditionals and operators to make the "Multiplication Station" app, which generates random, timed multiplication problems for you to solve and then check your answers.

Chapter 5: Beat the Bus: Tracking Location with Maps and Sensors Here you'll create and use lists to build the "Beat the Bus" app, which lets an approved adult track your ride from school to a preset destination, without using the power-draining location services on the adult's phone.

Chapter 6: Tic Tac Toe: Using Loops to Create a Game   You'll use loops, conditionals, and operators to create the classic "Tic Tac Toe" game.

Chapter 7: Multiplication Station II: Reusing Code with Procedures In this chapter, you'll explore how to create and call functions (called *procedures* in App Inventor) to build "Multiplication Station II," which lets you choose the difficulty level of the multiplication problems and computes both your raw and percentage scores.

Chapter 8: Virtual Shades: Drawing and Dragging Images   You'll create the "Virtual Shades" app, which lets you take a selfie, virtually try on a variety of sunglasses, and even doodle on the screen.

Appendix: App Inventor Components and Built-in Blocks

HOW TO USE THIS BOOK

To create the app in each chapter, you'll need to upload the images and any sound files. You can download them at *https://nostarch.com/programwith appinventor/*. To use your own images and sounds, use JPG or PNG image files and MP3 sound files in the smallest possible sizes. Keep in mind that App Inventor restricts the size of each uploaded file to 1MB, and no app may exceed 5MB (including pictures, sounds, and other assets).

Download the solutions to each chapter's "On Your Own" exercises and lesson plans for each chapter at *https://nostarch.com/programwithappinventor/*.

GETTING STARTED WITH APP INVENTOR

Before you can start designing mobile apps, you'll need to set up an App Inventor account. Having an account lets you work on your apps on the web and save them in the cloud. You'll also be able to *live-test* your projects, which means you can see and test all changes as you work.

In this section, you'll learn how to set up your own account and log in to App Inventor. Then you'll learn how to test your apps on a device.

To set up an account, visit App Inventor's home page at *http://appinventor .mit.edu/*, shown in Figure 1, on a web browser like Chrome or Firefox.

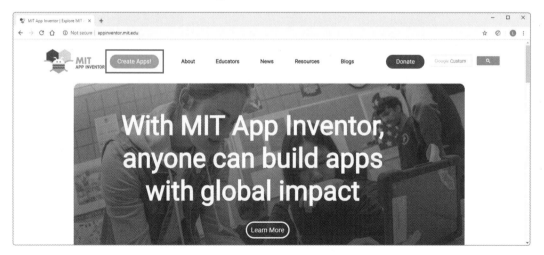

Figure 1: The App Inventor home page

Click the **Create apps!** button, outlined in red in Figure 1, at the top of the window. That should take you to a Google sign-in page like the one shown in Figure 2, where you'll be prompted to sign in with your Google account.

Google

Sign in
with your Google Account

Email or phone

Forgot email?

More options   NEXT

English (United States) ▼   Help   Privacy   Terms

Google

Sign in
with your Google Account

Email or phone

Forgot email?

Create account

Not your device?

NEXT

English (United States) ▼   Help   Privacy   Terms

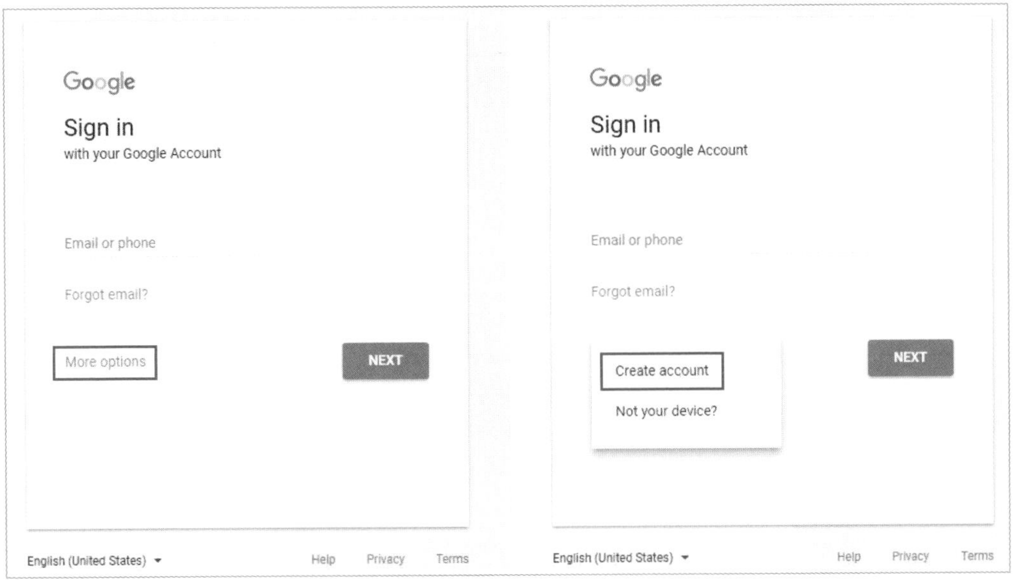

Figure 2: Sign in with your Google account

If you already have a Gmail account, enter that email address and, when prompted, enter your password. If not, create a Gmail account by clicking **More options** and then **Create account**. If you're under age 13, be sure to get permission and help from an adult to create the account.

Once you enter your Gmail address and password, you'll be asked to allow App Inventor to access your selected Google account, as shown in Figure 3.

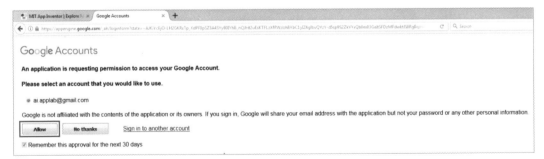

Figure 3: After you enter your Gmail address and password, App Inventor requests permission to access your Google account.

Click **Allow** to log in. If this is your first time logging in, you should see App Inventor's terms of service, as shown in Figure 4.

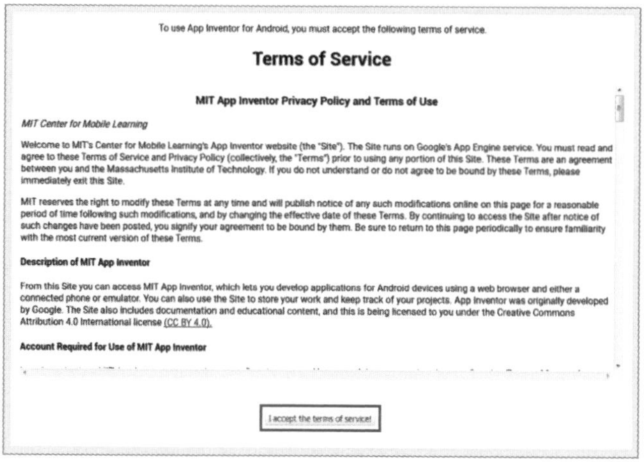

Figure 4: The terms of service window appears the first time you log in to App Inventor.

After you read them, click **I accept the terms of service!**.

Next you'll be asked to complete the voluntary user survey, as shown in Figure 5.

Figure 5: This survey window appears each time you log in until you take the survey or click the Never Take Survey button.

Click **Take Survey Now** to complete the survey, or click **Take Survey Later** to take it another time.

Now you should see the "Welcome to MIT App Inventor" screen in Figure 6, which displays each time you log in. This screen will let you know when the App Inventor developers are planning updates or releasing versions of App Inventor with fixes or new features.

Quickly read this screen for any updates, and click **Continue** to close the screen.

Figure 6: The welcome screen appears each time you log in to App Inventor and provides brief updates.

If this is your first time logging in to App Inventor, you should see the notice in Figure 7 indicating that you don't have any projects yet.

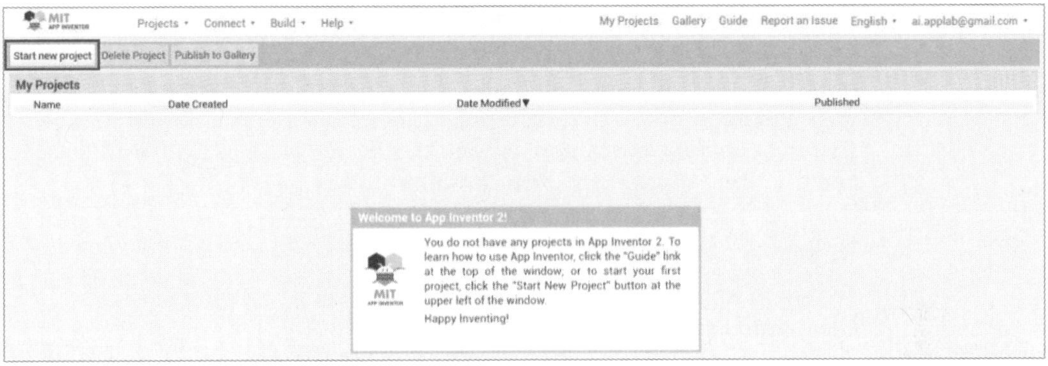

Figure 7: The Start new project screen appears when you log in for the first time.

To start your first project, click **Start new project** near the top-left corner of the window.

If you've logged in and saved a project before, you should see the last app you worked on at this point. To start a new app, click **Projects** to open the drop-down list and then select **Start new project**.

Now you're almost ready to create your first app!

LIVE-TESTING YOUR APPS

To create an app, you need to be able to see how it looks and works as you go. You'll find several options to test your code on the App Inventor website. For this book, we'll use App Inventor's Companion app, called *MIT AI2 Companion*, to connect your phone to your computer via Wi-Fi.

Although you can test App Inventor apps on all types of mobile Android devices, and testing with iOS devices is in development at the time of this writing, I recommend you use a full-featured phone. Using a full-featured phone ensures you'll have all of the modern mobile computing features (camera and video recorder; speech recognition, text-to-speech, and GPS capabilities; and a working texting plan) easily available to test in your apps.

CONNECTING YOUR PHONE TO THE COMPUTER

For live-testing to work properly, you must install the MIT AI2 Companion app on your device and connect your device to the internet through Wi-Fi. You don't need to download anything to your computer; simply follow the instructions in this section to download and install the MIT AI2 Companion app to the device you'll use for testing.

Installing the MIT AI2 Companion App

To download the MIT AI2 Companion app directly to your testing device, search for the app in your device's Google Play Store and then install it.

Once you've installed the Companion app, be sure that your device is using Wi-Fi (not the cell network) and that you've connected both the computer where you're logged in to App Inventor and your testing device to the *same* Wi-Fi network so that the two can communicate. You should now be able to open the Companion app on your device whenever you want to see and test your current project in App Inventor.

Testing with Your Device

To test an App Inventor project on your device, select **Connect ▸ AI Companion** as shown in Figure 8.

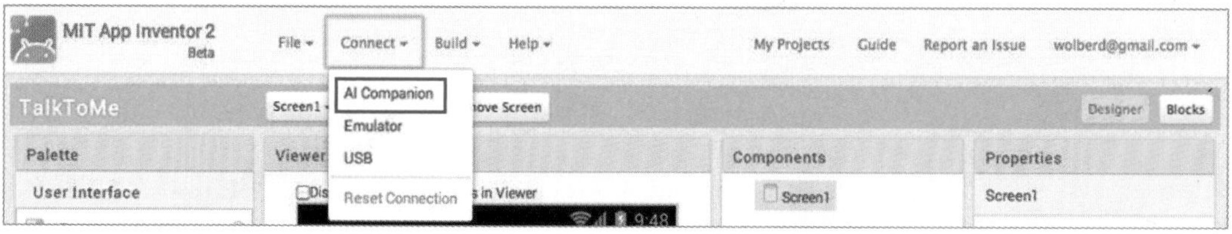

Figure 8: The Connect menu for live-testing apps, with the AI Companion option selected

You should see a small window displaying both a QR code and a six-character code, as shown on the left in Figure 9. When you open the Companion app on your device, it should look like the right side of Figure 9.

Figure 9: The App Inventor window that displays the code to scan or type into the AI2 Companion app to live-test on your device (left) and the AI2 Companion app's home screen (right)

Enter the six-character code in the Companion app's text box and click **connect with code**, or click **scan QR code** to scan the code. A few seconds later, the Companion app should display the project you're working on, which will automatically update as you make changes on your computer.

USING THE ONSCREEN EMULATOR

If you can't test with a device, you can test other ways, including using App Inventor's built-in Android emulator.

I will warn you now that the emulator works slowly and can't readily test certain mobile computing features, such as making calls, sending text messages, and responding to motion sensors. However, you'll still be able to check all aspects of your app's design and some features, like button clicks. To install the emulator, go to *http://appinventor.mit.edu/explore/ai2/setup-emulator.html* and then follow the instructions for your computer's operating system.

A NOTE ON CONVENTIONS

The book uses a number of conventions to streamline instructions or save space, as described next. Note that these methods may not always represent best coding practices.

HANDLING DUPLICATE COMPONENTS

A component added to an app is renamed with a more descriptive name only when two or more of the same components are used in an app, to avoid confusion. Otherwise, components retain their default names to limit non-essential instructions.

For example, if an app uses two or more buttons, we'll give them both meaningful names so we can clearly identify each button's function. But if an app has just one button, we'll leave it named Button1, since there's no possibility we'll confuse it with another button.

APPEARANCE OF BLOCK INPUTS

App Inventor lets you right-click blocks that require multiple inputs and select whether to display them with *inline inputs* or *external inputs*. The two random integer blocks in Figure 10 demonstrate the difference.

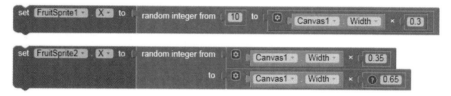

Figure 10: Setter blocks attached to two random integer blocks, the top with inline inputs and the bottom with external inputs

The FruitSprite1 setter block on the top shows blocks plugged into the from and to sockets in the random integer block as inline inputs.

But the FruitSprite2 block on the bottom shows blocks plugged into the from and to sockets as external inputs, taking up less horizontal space.

Throughout the book, when you notice that the inputs shown in a figure don't look exactly like those you're dragging into an app, it's because I've changed the input to better fit the available space on the page.

APPS CREATED ON WINDOWS

All the apps created in this book were created on a Windows computer. If you create your apps on a Mac, they may look different from those shown in the figures.

Once you've set up your App Inventor account and installed either the Companion app on your testing device or otherwise set up live-testing, you're ready to start creating and testing apps.

Let's get started!

1

BUILDING APPS WITH APP INVENTOR

With App Inventor, you can quickly and easily build apps using some of the most exciting mobile technology available. App Inventor provides tools that control the look and feel of an app, add useful mobile computing functions, and set or program an app's behavior.

In this chapter, you'll start to use these tools to build your first app, "Hi, World!", which lets you text someone using voice recognition. You'll also learn how to share your app with others.

PLANNING A NEW APP

Before you start building an app, you'll need to figure out the app's goal or what problem you want it to solve. Once you've done that, you can *decompose* your project, which means breaking down your goal into smaller parts that will be easier to tackle than trying to accomplish it all at once. This in turn will help you outline the precise steps, or *algorithm*, necessary to develop each part of the app before you start coding.

You'll also identify the elements and features the app will need to perform properly and allow users to interact with it as you intend. For example, will the app let users take videos, send text messages, or sense their locations? Will a user need to click a button, enter text, or select from a list of options to complete those tasks? These are the kinds of questions that might help you determine the features you'll need.

Once you know what to include in your app, you can lay out its *user interface* elements—the parts of the app that users can see and interact with on the screen. Your app should be easy for users to use and understand. To keep users' attention, you might add pictures, colors, and different text styles.

LAYING OUT YOUR APP

Once you have a plan, you'll start by laying out the necessary components in the Designer. In App Inventor, *components* are all the visible and non-visible elements you can use in your app, including those that set the look and feel and add exciting functionality. All components have *properties*, or characteristics, that you can set and/or actions you can program.

In this chapter, you'll use User Interface components as well as Media and Social components to add speech recognition and texting capability to your app. Later in this book, you'll use many more components to add features like video recording, location sensing, text-to-speech, animation, drawing, dragging, sound, picture taking, and more! For a detailed overview of available components, check out the appendix.

To lay out your app's components, you'll need to use App Inventor's Designer window, shown in Figure 1-1. The Designer window consists of a *Palette* pane to the left, which includes drawers of all possible components; a *Viewer* pane, where you can see the components you drag from the Palette and lay them out as they should appear on your app's screen; a *Components* pane, which lists all components in your app and allows you to give them meaningful names; a *Media* pane that lists all the pictures, sound clips, and other media files you've uploaded to the app; and a *Properties* pane, where you set the initial characteristics of app components.

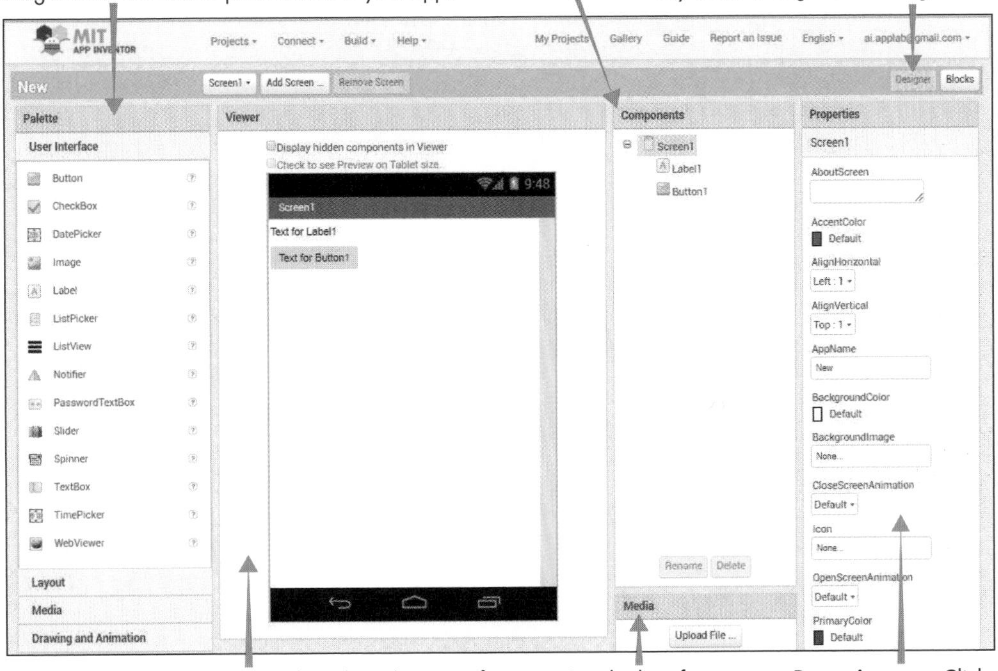
Components pane: See the list of components you've added to your app.
Palette pane: Look in the drawers for components and drag them to the Viewer pane to add to your app.
Designer button: Click this button from any window to go to the Designer.
Viewer pane: Drag components here from the Palette pane to add them to your app and see how they'll look on your app's screen.
Media pane: See the list of picture, sound, and other media files you've uploaded to your app.
Properties pane: Click a component and then change its properties here.

Figure 1-1: The App Inventor Designer window

PROGRAMMING YOUR APP IN THE BLOCKS EDITOR

Once you place all the necessary components in your app, use the Blocks Editor to program it according to your plan. Click the **Blocks** button shown in Figure 1-2 to open the Blocks Editor window.

The Blocks Editor window consists of the *Blocks* pane on the left and the *Viewer* pane on the right. The Blocks pane contains drawers for both *built-in blocks*, which program the app's general behaviors, and *component-specific blocks*, which direct particular actions for each app component. To tell your app what to do, drag blocks from the drawers in the Blocks pane and snap them together in the Viewer.

Viewer pane: Drag blocks from the Blocks pane drawers and snap them together here to program your app's behavior.

Blocks button: Click this button from any window to go to the Blocks Editor.

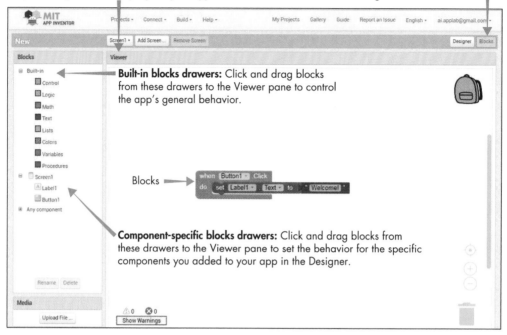

Built-in blocks drawers: Click and drag blocks from these drawers to the Viewer pane to control the app's general behavior.

Blocks

Component-specific blocks drawers: Click and drag blocks from these drawers to the Viewer pane to set the behavior for the specific components you added to your app in the Designer.

Figure 1-2: The Blocks Editor window

For a detailed overview of App Inventor's built-in blocks, check out the appendix. In later chapters, you'll explore component-specific blocks and the corresponding actions for a variety of components.

BUILDING THE "HI, WORLD!" APP

The first app you'll create in this book is the "Hi, World!" app. The app's name is a play on "Hello, World," a common first script that computer science students write when learning programming languages. The script displays "Hello, World" on a screen, but the "Hi, World!" app is so much more useful because it uses the SpeechRecognizer, PhoneNumberPicker, and Texting components to let you text using voice recognition! You can use this type-free texting app to say a message aloud, select a person's telephone number, and then text the person.

Log into App Inventor following the instructions outlined in "Getting Started with App Inventor" on page xviii. If this is your first time logging in, you should see the notice in Figure 1-3 letting you know that you don't have any projects yet.

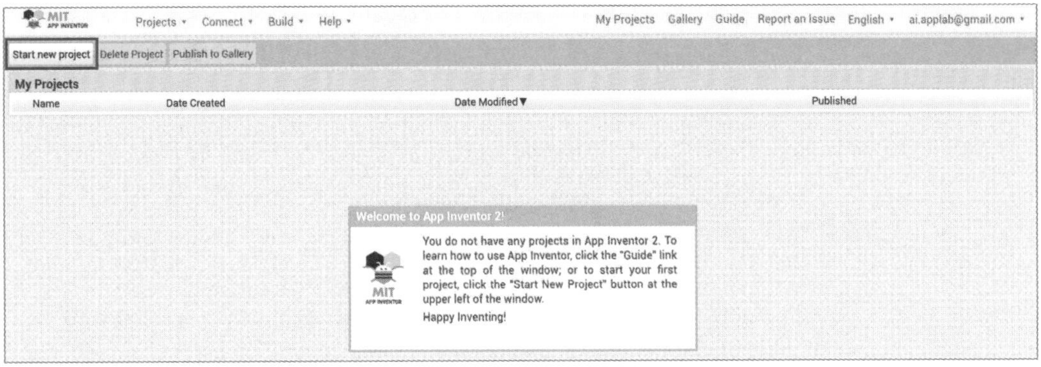

Figure 1-3: The Start new project screen appears the first time you log into App Inventor.

Click the **Start new project** button near the top-left corner of the window. Clicking the button should open the dialog where you'll enter the name for your first project. Enter HiWorld without any spaces, and click **OK**.

If you've logged in and saved a project before, App Inventor should open to the last app you worked on. To start a new app, click **Projects ▸ Start new project** as shown in Figure 1-4.

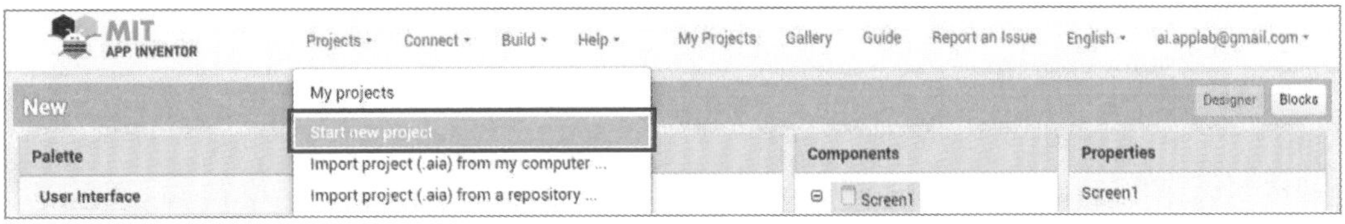

Figure 1-4: The Start new project option in the Projects menu tab

In the dialog that opens, name the project by entering HiWorld without any spaces and click **OK**.

DECOMPOSING "HI, WORLD!"

Before we start building the app, let's outline the actions we want "Hi, World!" to perform. We want the app to text someone a message that the user says aloud. We can break this action down into three steps.

1. When a user clicks the message button, open the phone's speech-to-text function and capture the user's text.

2. When a user clicks the PhoneNumberPicker, open the phone's contact list.

3. When a user clicks the text button, text the message converted by the speech-to-text function to the selected phone number.

You'll need the following components:

- **Button** (2) for the user to click to start the action
- **Image** for app design
- **PhoneNumberPicker** for the user to click to select a number
- **SpeechRecognizer** to capture and convert speech to text
- **Texting** to send the message

For each app we build throughout this book, we'll use this format to decompose the action into manageable steps and identify the components we need to make it work.

LAYING OUT "HI, WORLD!" IN THE DESIGNER

Now let's think about how to lay out the app's components and how to present instructions so the app is easy to use and understand.

For this app, the instructions to the user are so simple that we can include them on the Button and PhoneNumberPicker components, as long as those components are wide enough to display the text. To ensure that, we'll make the width of all visible components (those that the user can see and interact with) equal to the width of the screen. This means that we won't need to use a component from the Layout drawer to structure the screen display, because making everything the width of the screen requires us to place all components on top of each other. All we need to do is start with a picture at the top for visual effect and simply place the other components in the order we want users to interact with them.

Adding the Components

Go to the Designer to lay out the image and other components for all three parts of the app. Make sure you're in the Designer window by clicking the **Designer** button, as shown in Figure 1-5.

Figure 1-5: Click the **Designer** button to open the Designer window where you'll lay out the app components.

In the Designer window, drag the components you need from their drawers in the Palette pane to the Viewer pane. From the User Interface drawer, drag an **Image** and two **Button** components; from the Media drawer, drag a **SpeechRecognizer** component; and from the Social drawer, drag a **Phone NumberPicker** component and a **Texting** component. Your screen should now look like Figure 1-6.

In the Viewer pane, you should see all visible components on Screen1 ❶ and the non-visible components under the screen ❷. In the Components pane ❸, you should see a list of all components you just dragged to the Viewer pane.

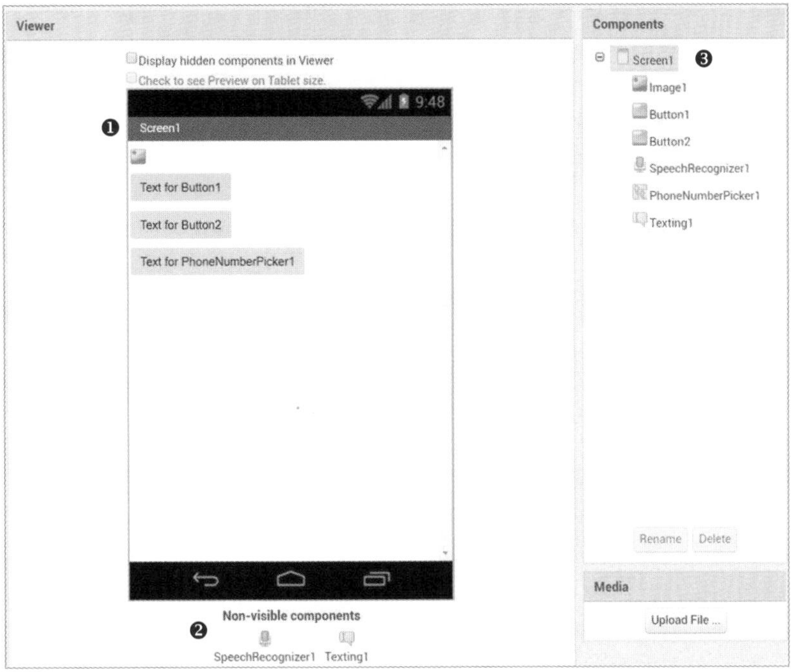

Figure 1-6: The screen after you drag all "Hi, World!" app components to the Viewer pane

Now you can make adjustments to each visible component by clicking it in the Components pane. Figure 1-7 shows what happens when you click Image1.

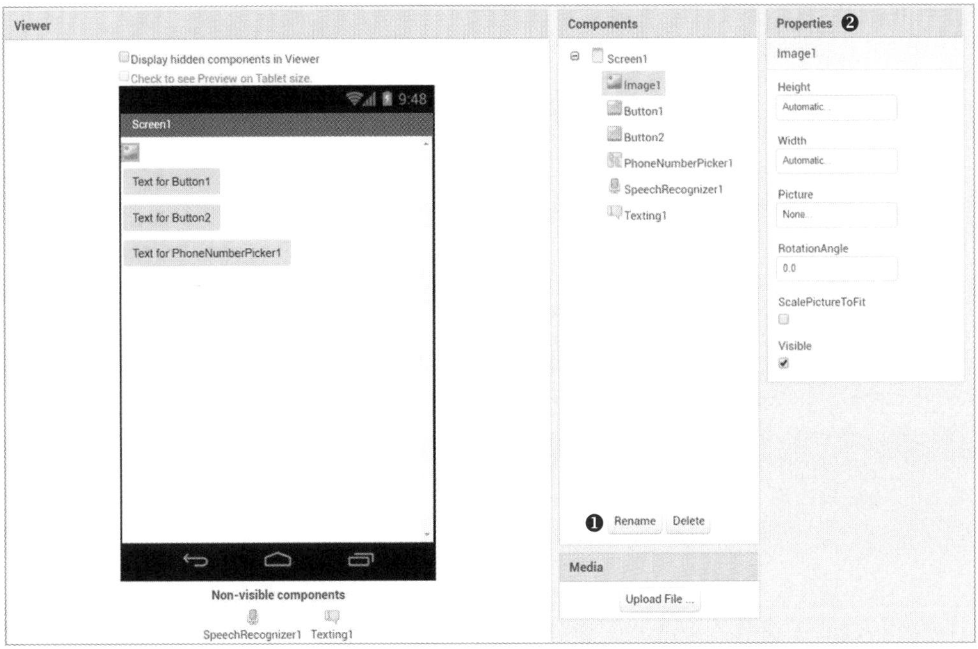

Figure 1-7: The Designer window after you click Image1 in the Components pane

Once you click Image1 in the Components pane, you should see buttons near the bottom of that pane ❶ that allow you to rename or delete the Image1 component. In the Properties pane ❷, you should see all the properties for Image1 that you can set in the Designer.

Resizing a Component's Width

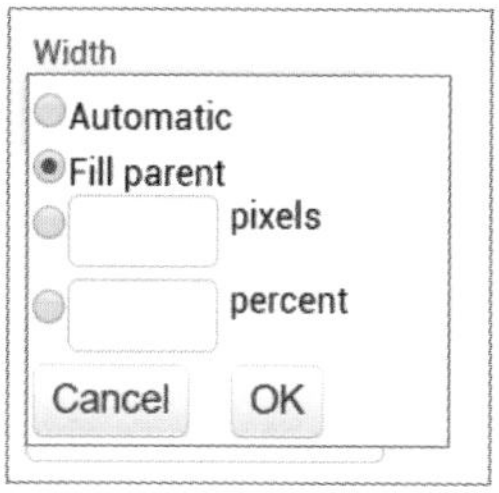

Since you have only one image in this app, you don't need to rename it. But you *do* need to change its size to make it the width of the screen. To resize Image1, in the Properties pane, click inside the text box under **Width** to open the dialog shown in Figure 1-8.

Click the **Fill parent** radio button and then click **OK**. This should make the width of Image1 fill the width of the screen, which is Image1's *parent*, or the component that contains it.

Figure 1-8: The Properties dialog that allows you to change a component's width

Uploading an Image

Next, you'll need to upload the picture you want to display as Image1 on the app screen. To upload your image, click inside the text box under **Picture** and click the **Upload File ...** button, shown on the left of Figure 1-9, which should open the dialog shown on the right.

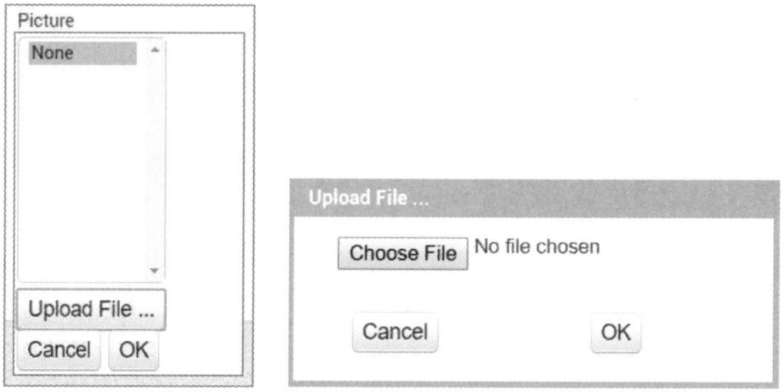

Figure 1-9: The Picture and Upload File ... dialogs

Click the **Choose File** button. The file manager should open to show the files on your computer and allow you to find the picture you want to upload, as shown in Figure 1-10.

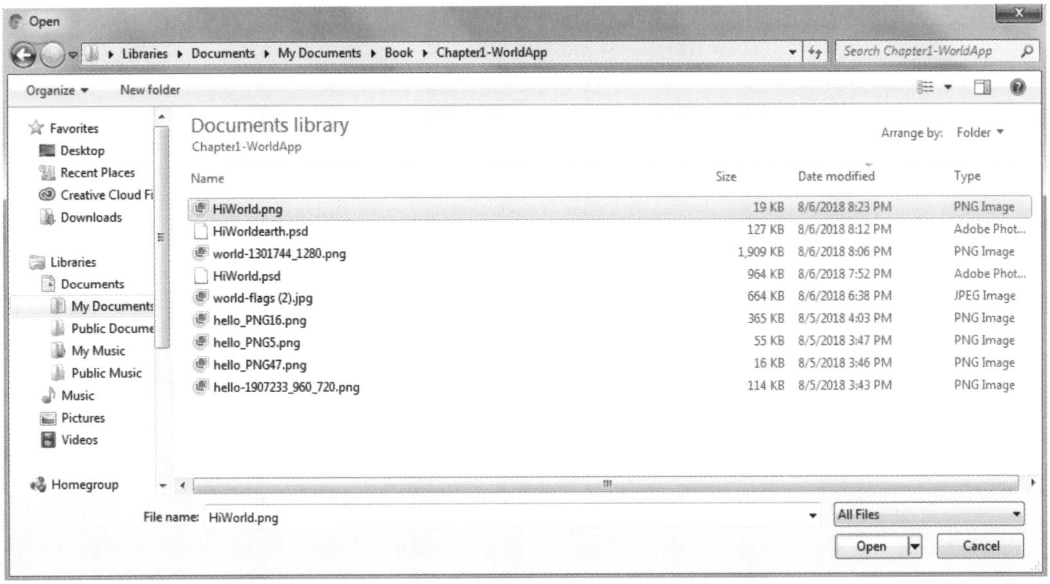

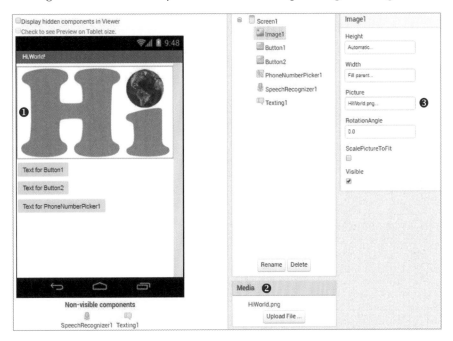

Figure 1-10: The file manager window allows you to select a file from your computer.

Select the image you want to use, click the **Open** button, and then click **OK** on the Upload File … dialog shown in Figure 1-9. You can download the sample image to use for this exercise, along with other materials that come with this book, at *https://nostarch.com/programwithappinventor/*.

Figure 1-11 shows what you should see after uploading the image.

Figure 1-11: The Designer window after you upload an image in the Properties pane

You should now see the name of the picture listed in the Picture text box for Image1 in the Properties pane ❸ and in the Media pane ❷. The uploaded picture should display on Screen1 in the Viewer pane ❶.

Creating the Say a Message Button

Now we'll set the properties for Button1 by clicking it in the Components pane. Since the app contains two Button components, let's change the name of Button1 to avoid confusion about each one's function.

To rename Button1, click the **Rename** button near the bottom of the Components pane. That should open a Rename Component dialog, as shown in Figure 1-12.

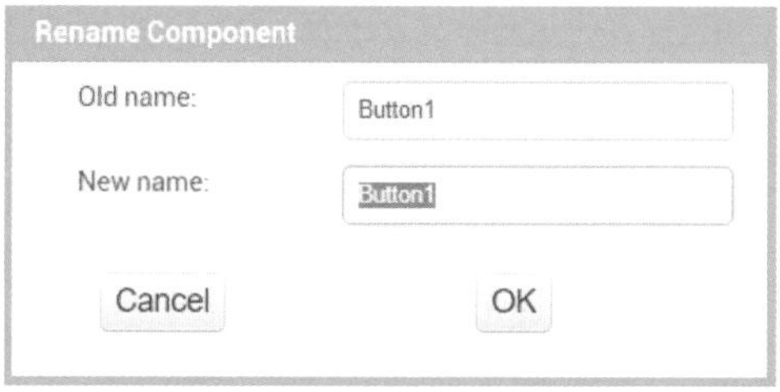

Figure 1-12: The Rename Component dialog opens after you click the Rename button in the Components pane.

NOTE   *When you rename a component, you change its default name to a more meaningful one that describes its function and is easily identifiable in the Designer window and Blocks Editor. Only you and the people you've shared your app code with will see a component's name. Renaming a component is different from changing the display text on a component like a Button or Label, which your users actually see.*

Enter **SayMessageBtn** without any spaces in the **New name:** text box and click **OK**. You should see that the Button name has changed to SayMessageBtn in the Components and Properties panes in the Designer window.

Let's make the width of the button fill the width of the screen as we did with Image1. To change the width of SayMessageBtn in the Properties pane, click inside the text box under **Width**, click the **Fill parent** radio button, and then click **OK**.

Finally, let's change the text to display on SayMessageBtn to provide clear instructions for the user. Click inside the text box under **Text** in the Properties pane. Delete the default Text for Button1 and type **1. Say a Message**, and then click outside of the text box. Figure 1-13 shows the new text on SayMessageBtn as well as the other changes you've made to its properties.

Figure 1-13: The Designer window after you change the name, width, and text of SayMessageBtn

Next, we'll make similar changes to Button2.

Creating the Text Your Message Button

Rename Button2 by clicking it and the **Rename** button in the Components pane, entering SendTextBtn without any spaces in the **New name:** text box, and then clicking **OK**. You can see that the button name has changed to SendTextBtn in the Components and Properties panes.

Next, to change the width of SendTextBtn in the Properties pane, click inside the text box under **Width**, click the **Fill parent** radio button, and then click **OK**. Finally, change the text on SendTextBtn to provide instructions for the app user.

To make the change, click inside the text box under **Text**, replace Text for Button2 with **3. Text Your Message**, and then click outside of the text box. In the Viewer pane, you should now see all the changes you made to SendTextBtn, as shown in Figure 1-14.

Figure 1-14: The Designer window after you change the name, width, and text of SendTextBtn

Next, let's update the app's final visible component, PhoneNumberPicker1, which the user will use to decide whom to text.

Creating the Select a Number Button

Click **PhoneNumberPicker1** in the Components pane. In the Properties pane, to change its width as you've done with the other components, click the **Fill parent** radio button and then click **OK**. Also change the text on the component to provide instructions for the app user.

To make the change, replace the existing Text for PhoneNumberPicker1 in the text box under **Text** with **2. Select a Number**, and then click outside of the text box. In the Viewer pane, you should see these changes to PhoneNumberPicker1.

Now all you need to do is move the PhoneNumberPicker component above SendTextBtn so that users will see the components in the order you want them to move through the app. We need the user to choose the recipient's phone number before sending the text, so we move PhoneNumberPicker1 up by clicking and dragging it up in the Viewer pane so that it comes before "3. Text Your Message". Figure 1-15 shows what the app looks like so far.

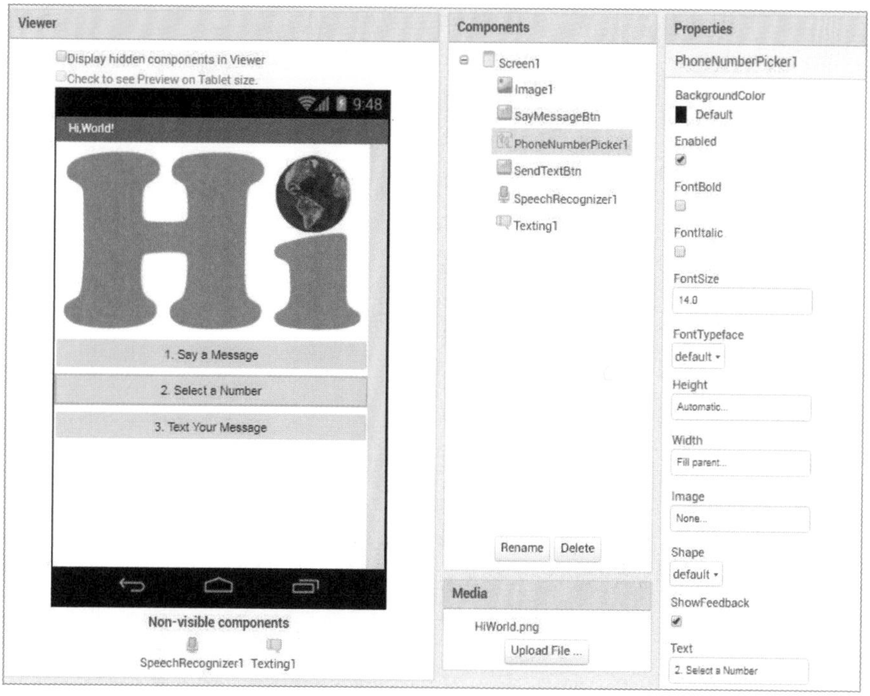

Figure 1-15: The Designer window after you change the width, text, and location of PhoneNumberPicker1

Try changing a few more properties to make the app even more visually appealing and user-friendly. For example, you can change the color of the Button components and PhoneNumberPicker1 to the darkest color in your image, such as the darker green in my Image1 component, and make the text bold, larger, and white so it's easy to see on the new dark background. Let's try that now.

Changing Button Color

To change the background color of the Button and PhoneNumberPicker1 components, click each component in the Properties pane and then click the word **Default** under BackgroundColor to open the color list dialog shown to the left in Figure 1-16.

If one of the colors in the list matches the darkest color in Image1, click the name of that color to select it. If none of the colors in the list work, click **Custom...** at the bottom of the list, which will open the color picker shown on the right in Figure 1-16.

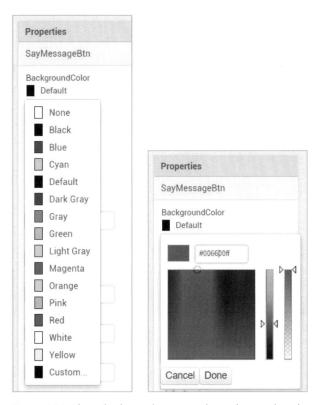

Figure 1-16: The color list and custom color picker used to change colors of various component properties in the Properties pane

In the color picker, replace the current 000000 color number by entering the number of the color you want to use. For instance, you can enter 006600 for the darker green in my Image1 component, as I've done here, and then click **Done**.

Formatting Button Font

Now make the text on each of the components bold by clicking the **FontBold** checkbox in the Properties pane, as shown in Figure 1-17.

Make the text larger by replacing 14.0 with 20 in the **FontSize** text box. Finally, you can make the text color white by clicking the word **Default** under TextColor near the bottom of the Properties pane to open the color list, as shown in Figure 1-17, and then clicking **White**.

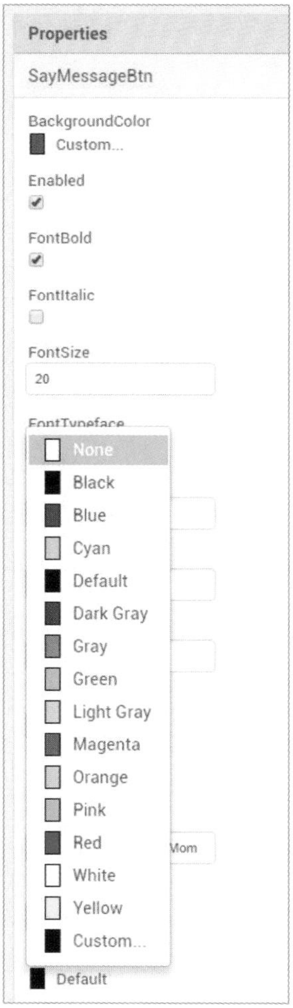

Properties

SayMessageBtn

BackgroundColor
■ Custom...

Enabled
☑

FontBold
☑

FontItalic
☐

FontSize
20

FontTypeface

☐ None
■ Black
■ Blue
☐ Cyan
■ Default
■ Dark Gray
■ Gray
■ Green
☐ Light Gray
■ Magenta
☐ Orange
■ Pink
■ Red
☐ White
☐ Yellow
■ Custom...

Mom

■ Default

Figure 1-17: The SayMessageBtn *Properties pane, showing changes to the* Button *background and font*

PROGRAMMING "HI, WORLD!" IN THE BLOCKS EDITOR

Now that you've laid out all the components, you can move to the Blocks Editor to program the app. Click the **Blocks** button, as shown in Figure 1-18, to switch to the Blocks Editor.

Figure 1-18: The Designer window after you lay out the app

For the "Hi, World!" app, you'll use only component-specific blocks to program the action. You'll program in the order of steps outlined in "Decomposing 'Hi, World!'" on page 5.

Step 1: Converting from Speech to Text

We start by telling the app what to do when the user clicks SayMessageBtn, the "1. Say a Message" button. When that button is clicked, we want the app to open the phone's speech-to-text feature to capture and convert the user's speech to text.

Figure 1-19 shows how to program step 1.

In the Blocks pane, click **SayMessageBtn** ❶ and, when the blocks for the component appear, drag the **whenSayMessageBtn.Click** block to the Viewer. Then, in the Blocks pane, click **SpeechRecognizer1** ❷ and drag the **callSpeech Recognizer1.GetText** block to the Viewer. Snap the **callSpeechRecognizer1.GetText** block inside the **whenSayMessageBtn.Click** block next to the word do. In plain English, these blocks say, "when the button is clicked, open the phone's speech recognizer to get text," which is very similar to what we outlined when we decomposed the app.

Now we'll live-test to see how these blocks work. It's best to test this app on a working phone. If you try to test these blocks with the built-in emulator or with a device that doesn't have speech-to-text capability, you'll likely get an error when you click **SayMessageBtn**. Use the MIT AI2 Companion app on a phone, as outlined in "Live-Testing Your Apps" on page xxii.

Figure 1-19: The Blocks Editor window with code blocks dragged onto the Viewer to program SayMessageBtn

Once you click **Connect ▶ AI Companion** in the top menu bar and scan the QR code with your phone's Companion app, your "Hi, World!" app should appear on your phone.

Click the **1. Say a Message** button. The phone's speech recognizer should open and prompt you to speak, as long as your phone's speech-to-text capability is turned on and the blocks are laid out as shown in Figure 1-19. So far, so good. Leave the app open on your phone so you can keep live-testing.

Step 2: Opening the User's Contact List

Now let's program the second part of the app, telling it what to do when the user clicks the "2. Select a Number" button. When the PhoneNumberPicker is clicked, we want the app to open the user's contact list for selection. Here's the code for programming this:

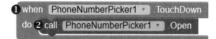

In the Blocks pane, click `PhoneNumberPicker1`, and when the blocks for the component appear, drag the `whenPhoneNumberPicker1.TouchDown` block ❶ to the Viewer. Then, in the Blocks pane, click `PhoneNumberPicker1` again and drag the `callPhoneNumberPicker1.Open` block ❷ to the Viewer and snap it inside the

whenPhoneNumberPicker1.TouchDown block next to the word do. In plain English, these blocks tell the app, "open the phone's contact list when the phone number picker is touched," like we planned.

Now live-test to see how these blocks work. Click **2. Select a Number**. Your phone's contact list and a prompt to select a contact should appear, assuming you have contacts saved in your phone.

Step 3: Sending the Text Message

Now we'll program the final part of the app so it knows what to do when a user clicks the "3. Text Your Message" button. That's when we want the app to set the message for the Texting component to the text converted by the device's speech-to-text function, set the phone number for the Texting component to the user's selected phone number, and then text the message to that phone number. Here are the blocks for SendTextBtn.

In the Blocks pane, click **SendTextBtn**, and when the blocks for the component appear, drag the **whenSendTextBtn.Click** block ❶ to the Viewer. Then, in the Blocks pane, click the **Texting1** component and drag the **setTexting1 .Messageto** block ❷ to the Viewer and snap it inside the **whenSendTextBtn.Click** block next to the word do.

Then, in the Blocks pane, click the **SpeechRecognizer1** component and drag the **SpeechRecognizer1.Result** block ❸ to the Viewer and snap it to the right side of the **setTexting1.Messageto** block.

Getters and Setters

To understand what we're doing with these blocks, you first need to understand that App Inventor's Texting component can send a text message only if its Message and PhoneNumber properties are set. That simply means that Texting1 must have a message to send (its Message property) and must know where to send it (its PhoneNumber property).

We tell the app to set the Message property with the SpeechRecognizer1 .Result block, which is called a *getter block* since it gets a value for us, and the setTexting1.Messageto block, which is called a *setter block* because it sets a value.

Technically, when we snap blocks together side by side as we've done here, they *execute*—or work—from right to left. That means that, with these blocks, the app first gets the current value of the SpeechRecognizer1 Result property, which is the text that we get from the app using the device's speech-to-text function. Then the app sets the Texting1 Message property to that text, which lets the Texting component know the message to send.

As with the properties for many other components, you can set the Texting *component's* Message *and* PhoneNumber *properties initially in the Designer window's Properties pane if it makes sense for your app, or you can have the app set or change them dynamically using component-specific blocks like the ones we're using here. For instance, if you wanted your app to send the exact same message each time, it would make sense to set the* Texting *component's* Message *property by typing it in the Properties pane.*

Let's continue programming. In the Blocks pane, click the **Texting1** component again, and drag the **setTexting1.PhoneNumberto** block ❹ to the Viewer and snap it under the **setTexting1.Messageto** block. Then, click **Phone NumberPicker1** in the Blocks pane and drag the **PhoneNumberPicker1.PhoneNumber** block ❺ and snap it to the right side of the **setTexting1.PhoneNumberto** block.

Here, when the user clicks SendTextBtn, we have the app set the other property required for Texting1—the PhoneNumber property—with the getter block PhoneNumberPicker1.PhoneNumber and the setter block setTexting1.Phone Numberto. The app first gets the current value of the PhoneNumberPicker1 Phone Number property, which is the phone number the user selected when the app opened the device's contact list, and then sets the Texting1 PhoneNumber property to that number. This lets the Texting component know the phone number to send the text to.

Finally, click **Texting1** in the Blocks pane again, drag the **callTexting1 .SendMessageDirect** block ❻ to the Viewer, and snap it under the **setTexting1 .PhoneNumberto** block. Now, when the user clicks SendTextBtn, after the app sets the necessary properties for Texting1, this block directs the app to send the message directly.

Instead of sending the message directly, we could have the app open the default texting app so the user can edit the message before sending. To do this, use the **callTexting1.SendMessage** block in place of the **callTexting1 .SendMessageDirect** block.

In plain English, all together your blocks simply say, "when the button is clicked ❶, set the text message for the Texting component ❷ to the text converted by the speech recognizer ❸, set the phone number for the Texting component ❹ to the phone number selected in contacts ❺, and then send the message ❻," just like we planned.

Now live-test the completed app! On your phone, click **1. Say a Message** and say a message when the speech recognizer prompts you. Then, click **2. Select a Number** and select a phone number from your contact list. Finally, click **3. Text Your Message** and then check whether your phone sent the text. As long as you're testing on a working phone with a texting plan, if you placed your blocks as shown in the code examples, the app should work correctly.

If the app doesn't work, you now must *debug*, which means to look closely at your code for *bugs* (errors) and fix them. Recheck your blocks to make sure you've dragged the getter and setter blocks into the right spots, and test again after you make any corrections. Before retesting, to make sure your changes take effect, reload the app in the Companion by changing any property in the Property pane, like clicking any checkbox. Once all the buttons work, you have successfully created your first fully functional app!

SHARING "HI, WORLD!"

You can email your app for others to install on their phones. But first, click **Screen1** in the components pane, so you can make sure the app's name displays the way you want it to when you send it to your friends. In the Properties pane, in the text box under **AppName**, you can see that the name now shows as *HiWorld*, which is the name you entered when you started the project. To change the app's name, click inside the text box, add a space between *Hi* and *World*, maybe add an exclamation point after *World*, and then click outside of the text box. Also, click the checkbox under **TitleVisible** to remove the check mark, which will keep the title of the screen, *Screen1*, from showing at the top of the app when it runs on a phone.

Now, with the app project open on your computer screen, click **Build ▶ App (save .apk to my computer)** in the top menu bar, as shown in Figure 1-20.

Figure 1-20: The Build menu link creates a file you can share with friends so they can install the app.

You should see a progress bar showing that the file is being saved. Once the file is saved, you can locate it on your computer and email it as an attachment to whomever you want. When people open the email on their phones, they'll be prompted to install the app. Note that since "Hi, World!" is a nonmarket app, your friends first will need to make sure their phones' settings allow installation of applications from "unknown sources."

SUMMARY

In this chapter, you built your first app, "Hi, World!", which converts a spoken message into a text message and sends it to a selected contact! In the process, you explored the app design process, which included learning how to decompose a big plan into its smaller steps, then following those steps to program the actual components. Now you should feel more comfortable navigating around the App Inventor Designer window and Blocks Editor.

In the next chapter, you'll examine how App Inventor uses event-driven programming to set app behavior, and you'll become familiar with events and event handlers. We'll use several examples of both to create the "Practice Makes Perfect" app, where users can record, watch, and re-record temporary video clips of themselves practicing speeches or songs.

Now that you've created "Hi, World!", save new versions of it as you modify and extend it while working on these exercises. You can find solutions online at *https://nostarch.com/programwithappinventor/*.

1. Change the app so that it prompts the user to enter a telephone number rather than selecting one from the contact list. Which component(s) would you drag from the User Interface drawer to the Viewer to enable the user to enter the telephone number manually? Now that you'll be replacing the PhoneNumberPicker component, how will you include instructions to the user about where to enter a phone number? What blocks will you use to provide the phone number required for Texting1?

2. Extend the app for Android devices so that, after it texts the spoken message, it waits for a response text message from the recipient and then reads that message to the user aloud.

3. Change the app so that it emails the message instead of texting it. What components and blocks would you use to send an email message?

2

APP INVENTOR AND
EVENT-DRIVEN PROGRAMMING

App Inventor apps use a model called *event-driven programming*, where we program behaviors to respond to certain events. An *event* can be something the user does or something that happens within or to the device.

As App Inventor programmers, we use *event handlers* called when...do blocks to tell the app how to respond to an event. For each when...do block, we can include one or more commands for the app to follow in sequence once the event occurs. Those commands include blocks to get and set values as well as call blocks to start built-in *methods* or our own written *procedures*, which are a series of defined tasks for the app to execute.

In fact, you already used event handlers to create the "Hi, World!" app in Chapter 1. Let's review all the blocks you used for "Hi, World!".

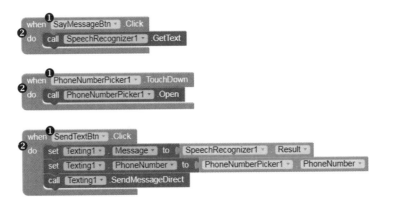

We can see the app's three events ❶, which are the two button *clicks* on SayMessageBtn and SendTextBtn and the *touch down* event on PhoneNumberPicker1. You used three when...do event handlers ❷ to program the actions the app should take in response to those events, such as calling the SpeechRecognizer method to get text from speech, calling the PhoneNumberPicker.Open method to open the phone's contact list, and setting the Texting component's required properties before calling the method that sends a text.

Although the user triggered all the events in "Hi, World!", that's not always the case. Some events are caused when something happens to the device. For instance, when a phone receives a call, you can use an event handler to deal with that event. You also can use event handlers to respond to automatic events or events that happen to screens or components in your apps. For instance, you can add blocks to program actions that should occur when the screen of an app first opens.

We'll play with several event handlers as we build the next app.

BUILDING THE "PRACTICE MAKES PERFECT" APP

In this chapter, you'll create the "Practice Makes Perfect" app, which uses the Clock, Camcorder, and VideoPlayer components to let users repeatedly record videos of themselves practicing speeches or songs and then watch the videos to review their performance. For convenience, we'll program "Practice Makes Perfect" to automatically open the device's video camera, getting it ready to record right away. That's different from many camera apps, which require the user to open the app, switch to video mode, and then record.

We'll program six event handlers for "Practice Makes Perfect," one to direct the action after a timer goes off five seconds after the app opens, another that tells the app what to do after the video camera records, and four that respond to the user's button clicks.

Let's get started! Log into App Inventor following the instructions outlined in "Getting Started with App Inventor" on page xviii. App

Inventor should open to the last app you worked on. To start a new app, click **Projects ▸ Start new project** to open the dialog where you enter the project name, enter `PracticeMakesPerfect` without any spaces, and click **OK**.

DECOMPOSING "PRACTICE MAKES PERFECT"

"Practice Makes Perfect" will automatically open the device's video camera to record a video and then play the video back on the app's own video player. After that, the app will let users record and view as many additional videos as they want. The app will also display a decorative temporary, or *splash*, title screen when the user first opens it.

We can break this action down into four steps:

1. When a user opens the app, display the splash screen for 5 seconds before displaying the Record button, Watch button, and the VideoPlayer. Open the device's video recorder so the user can record.
2. When the user clicks the Record button, open the device's video camera so the user can record.
3. Once the user records a video, set the recorded clip to play the next time the app's VideoPlayer starts.
4. When the user clicks the Watch button, play the user's video clip in the app's VideoPlayer.

You'll need the following components:

*   `Button` (2) for the user to click to manually start action
*   `Camcorder` to open the app's video camera
*   `Clock` to automatically start the action after a set interval
*   `Image` displaying the temporary splash graphic
*   `VerticalScrollArrangement` to hold the app's `Button` and `VideoPlayer` components
*   `VideoPlayer` to play video clips within the app

LAYING OUT "PRACTICE MAKES PERFECT" IN THE DESIGNER

Let's begin by laying out all the components, keeping in mind that we want the app to start with a separate splash screen and then automatically open the video camera to record the first video clip. We also want to arrange the components so that the app's easy to use and understand.

Adding the Components

Go to the Designer window and drag the listed components from their drawers in the Palette pane onto the Viewer pane. From the User Interface drawer, drag an `Image` and two `Button` components; from the Layout drawer, drag a **VerticalScrollArrangement**; from the Media drawer, drag a **VideoPlayer** and a **Camcorder** component; and from the Sensors drawer, drag a **Clock** component. Your screen should now look like Figure 2-1.

Figure 2-1: The screen after you drag the "Practice Makes Perfect" app components into the Viewer pane

In the Viewer pane, all visible components should show on Screen1 ❶, and the non-visible components (Camcorder and Clock) should show below the Screen ❷. In the Components pane ❸, you should see a list of all the components you've dragged onto the Viewer pane.

Creating a Splash Screen

As mentioned, we'll display a full-screen introductory graphic that creatively illustrates the app's name and purpose, but we'll show it only temporarily to keep it from taking up too much screen space. We'll use Image1 as the splash screen, and we need to adjust several properties to get it to display as desired.

First, click Image1 in the Components pane, and set its height and width to fill the screen by clicking the text boxes under both **Height** and **Width** in the Properties pane. When you click each text box, you should see the dialog shown in Figure 2-2.

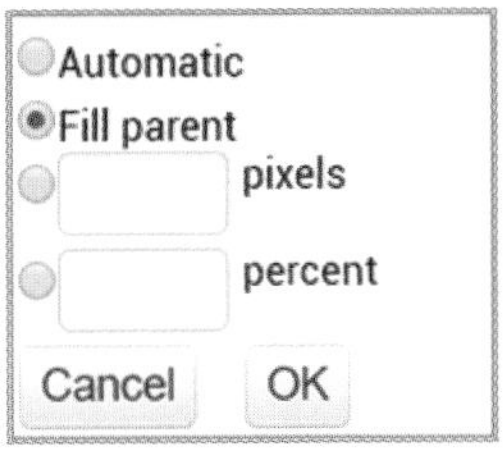

Figure 2-2: The Properties
dialog that allows you to
change a component's
height or width

Click the **Fill parent** radio button and then click **OK**. This will make
the height and width of Image1 take up the entire height and width of the
screen once the picture is uploaded.

Uploading a Picture

We now need to upload the picture we want to display as Image1 on the app
screen. To upload a picture, click the text box under **Picture** and click the
Upload File … button shown on the left in Figure 2-3, which should open
the dialog shown on the right.

Figure 2-3: The Picture and Upload File … dialogs

Click the **Choose File** button. The file manager should open to show
the files on your computer and allow you to find the picture you want to
upload, as shown in Figure 2-4.

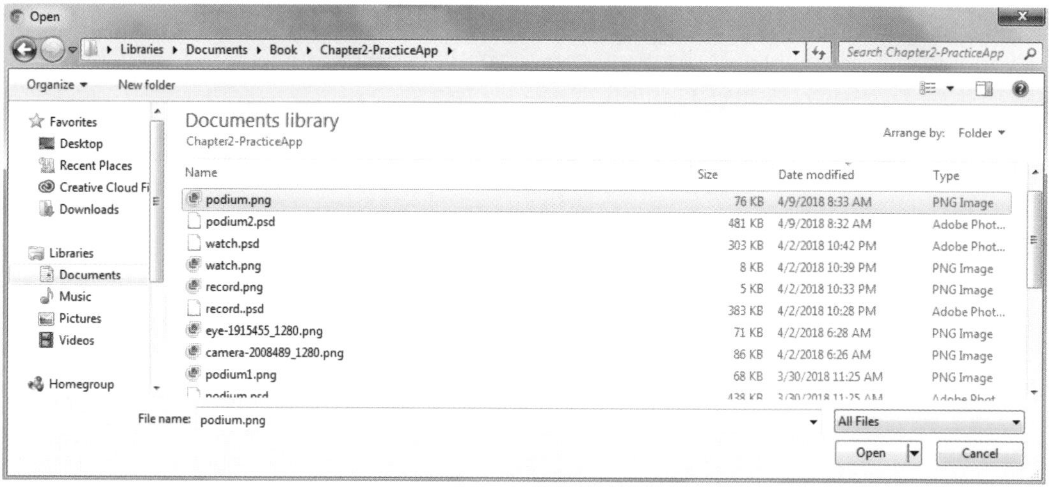

Figure 2-4: The file manager window allows you to select a file from your computer.

Click the image you want to use, click the **Open** button, and then click **OK** on the Upload File … dialog shown on the right in Figure 2-3.

NOTE *For some of the apps in this book, I created the decorative images we're using by searching the web for pictures, saving them, and then adding text to them using a graphics editor. Before saving any images from the web, however, I made sure I had the license, or right, to use and modify them, and I checked whether the license requires attribution, which means I'd need to give credit to the original artist when using the pictures. If you use images or other files you find on the web, always make sure that they are labeled as free to use in the way you intend and that you give credit to the creator if directed.*

You should now see the picture on Screen1 in the Viewer pane. Its width fills the screen, just as we want, but its height doesn't because the other visible components are taking up most of the screen's vertical space. We'll adjust this later when we make those components invisible.

You'll also notice in the Properties pane that, by default, the checkbox under the Image1 Visible property is checked, meaning users should see Image1 when the app opens. But because we want Image1 to show only temporarily, soon we'll program the app to change the value of the Image1 Visible property to hide the splash screen when we no longer want it to show.

Finally, while in the Image1 Properties pane, click the checkbox under ScalePicturetoFit. This will distort Image1 for now, making it take up even more of the screen width, but later, when Image1 is the only visible component on the screen as planned, you'll see that this is the effect we want.

Creating the Record and Watch Buttons

Next, let's set the properties for both Button components. To avoid confusion, let's give each Button a name that describes its function in the app, as we did in Chapter 1.

Click Button1 in the Components pane and then click the **Rename** button near the bottom of the pane. Replace Button1 with RecordBtn in the **New name:** text box and click **OK.** You should now see RecordBtn in the Components and Properties panes in the Designer window. Follow the same steps to change the name of Button2 to WatchBtn.

Using Images as Buttons

Let's also change the look of the buttons. For that, we'll use images with text and icons instead of the default App Inventor buttons. Click the text box under **Image** in the Properties pane for each button. Then upload the picture you want to use by following the steps just described in "Uploading a Picture" on page 27.

Once you upload the button images, be sure to delete the default text that's displayed on the buttons. To do so, click the text box under **Text** in the Properties pane for each button, delete the existing text, and then remove your cursor from the text box.

Finally, let's make each Button component and the VideoPlayer the width of the screen by clicking the text box under **Width** in the Properties pane for each component, clicking the **Fill parent** radio button and then clicking **OK.**

Grouping Components Vertically

As outlined earlier, step 1 of this app requires the Button components and VideoPlayer to be invisible for 5 seconds after the app opens. We can accomplish this by using the VerticalScrollArrangement component from the Layout drawer. Layout components allow you not only to uniformly align other components but also to combine components so you can program them as a group.

In this app, we're grouping all the components we want to hide temporarily when the app opens—RecordBtn, WatchBtn, and VideoPlayer1—inside a VerticalScrollArrangement. We're using the VerticalScrollArrangement in particular because we want to stack the components on top of each other, and we want to allow users to scroll vertically if necessary to see the videos in VideoPlayer1.

To group RecordBtn, WatchBtn, and VideoPlayer1 inside VerticalScroll Arrangement1 in that order, click each component in the Viewer pane, and drag it into **VerticalScrollArrangement1.** You should then see the components *nested within*, or contained inside, VerticalScrollArrangement1 in the Viewer and in the Components pane, as shown in Figure 2-5.

Figure 2-5: The Designer window after you drag RecordBtn, WatchBtn, *and* VideoPlayer1
inside VerticalScrollArrangement1

Let's now change the Visible property of VerticalScrollArrangement1. Click that component in the Components pane, and then click the checkbox in the Properties pane under **Visible** to remove the check mark. This should make VerticalScrollArrangement1 and all its contents invisible in the Viewer now and when the app starts. When we program the app, we'll have it adjust this setting while it's running to display these components 5 seconds after the app starts. Also, you'll see that because VerticalScrollArrangement1 is invisible, Image1 now fills the screen as planned.

The last component we need to adjust before we can start programming is the Clock. Click the **Clock** component in the Components pane to change its TimerInterval property, as shown in Figure 2-6.

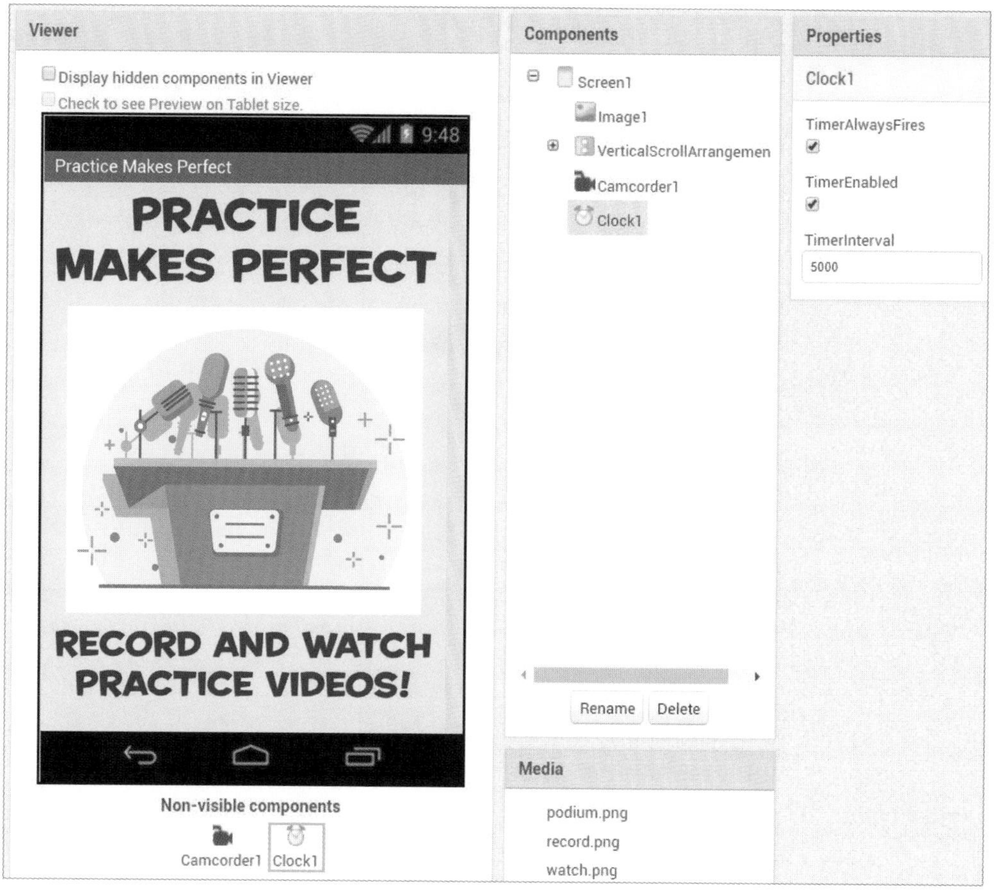

Figure 2-6: The Designer window after you lay out "Practice Makes Perfect," showing the adjusted property pane for Clock1

To make the change, click the text box under **TimerInterval**, replace the default value of 1000 with **5000**, and then click your cursor outside of the text box. This sets the Clock component's timer interval to 5,000 milliseconds, or 5 seconds. This means that, as long as the Clock component's timer is enabled and set to *fire*, or go off, as in Figure 2-6, it will automatically do what we program it to do every 5 seconds. But for your app, you'll need to adjust the Clock1 properties while the app's running so the timer fires just once to make Image1 disappear after 5 seconds when the Button, VideoPlayer, and Camcorder components appear, instead of firing every 5 seconds without stopping.

PROGRAMMING "PRACTICE MAKES PERFECT" IN THE BLOCKS EDITOR

Now that you've laid out all the components, you can move to the Blocks Editor to program the app. Click the **Blocks** button to switch to the Blocks Editor, and let's begin programming the four steps in order.

STEP 1: STARTING THE APP

We start by telling the app what to do when it opens with the splash screen, Image1, displayed. We want it to wait 5 seconds before hiding Image1, showing the Button and VideoPlayer components for the remaining time that the app is open, and opening the Camcorder.

This code shows how to set up this first step using the Clock component's *timer* event handler.

In the Blocks pane, click **Clock1**, and when the blocks for the component appear, drag the **whenClock1.Timer** event handler block ❶ to the Viewer. Next, in the Blocks pane, click **Image1** and drag the **setImage1.Visibleto** block ❷ to the Viewer and snap it inside the **whenClock1.Timer** block next to the word do. Then, return to the Blocks pane, click the **Logic** blocks drawer, and drag the **false** block ❸ to the Viewer and snap it to the right side of the setImage1 .Visibleto block. These are the blocks that set the value of the Image1 Visible property to false after the Clock component's timer fires, which hides the Image1 splash screen after the 5-second time interval we set for the Clock in the Designer.

NOTE *In the Logic blocks drawer, App Inventor provides two Boolean value blocks, one with a value of true and the other with a value of false. Boolean variables and properties—such as the Visible property of most visible components and the Clock component's TimerEnabled property—have only two possible values, true and false. You can set the initial values of these properties in the Properties pane by checking or unchecking the property's checkbox, and you can change the values with the blocks while the app's running.*

Now that we've programmed the splash screen to disappear after 5 seconds, we need to make the buttons and video player visible. To do this, click **VerticalScrollArrangement1**, drag the **setVerticalScrollArrangement1.Visibleto** block ❹ to the Viewer, and snap it inside the **whenClock1.Timer** block under the setImage1.Visibleto block. Then, in the Blocks pane, click the **Logic** blocks drawer, and drag the **true** block ❺ to the Viewer and snap it to the right side of the setVerticalScrollArrangement1.Visibleto block. These blocks set the value of the VerticalScrollArrangement1 Visible property to true after the Clock component's timer fires, which makes VerticalScrollArrangement1 and everything inside of it visible after the 5-second time interval we set for the Clock in the Designer.

Next, in the Blocks pane, click **Camcorder1**, drag the **callCamcorder1** .RecordVideo block ❻ to the Viewer, and snap it inside the **whenClock1.Timer** block under the `setVerticalScrollArrangement1.Visibleto` block. These blocks call the `Camcorder` component's built-in `RecordVideo` method to automatically open the user's video camera after the `Clock`'s 5-second time interval.

Finally, click **Clock1**, scroll down to find the **setClock1.TimerEnabledto** block ❼, drag it to the Viewer, and snap it inside the **whenClock1.Timer** block under the `callCamcorder1.RecordVideo` block. Then, in the Blocks pane, click the **Logic** blocks drawer, and drag the **false** block ❽ to the Viewer and snap it to the right side of the `setClock1.TimerEnabledto` block. These blocks set the value of the `TimerEnabled` property for `Clock1` to `false`, which stops the `Clock1` timer. This means that the timer will fire only once to execute the commands in this event handler after the first 5 seconds the app is open, and will not fire again until the app reopens.

In plain English, for step 1, you set the Designer properties to display the splash screen (`Image1`) when the app opens. Then, after 5 seconds, the code blocks tell the app to hide the image, display the buttons and video player, and open the video recorder to record a video.

Now live-test to see how these blocks work, preferably with your phone using the MIT AI2 Companion app, as outlined in "Live-Testing Your Apps" on page xxii.

Once you click **Connect ▸ AI Companion** in the top menu bar and scan the QR code with your phone's Companion app, your "Practice Makes Perfect" app should open on your phone. As long as your blocks are placed as shown in the code example, you should see `Image1` for 5 seconds before the buttons and video player appear and the video camera opens. You should not see the buttons and video player until you close the video camera. If you close the video camera without recording, you'll likely see an error saying you didn't record a video, which is true and fine for now. Leave the app open on your phone so you can keep live-testing.

STEP 2: OPENING THE VIDEO RECORDER

So far we've programmed what the app should do when the user first opens it. Now let's program step 2 of the app, which lets the user manually open the video camera to record a video when needed.

Here's where we tell the app what to do when users click `RecordBtn`, which is the way users reopen the video camera after it closes. When `RecordBtn` is clicked, we want the app to open the device's video camera so the user can record.

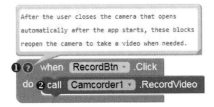

In the Blocks pane, click **RecordBtn**, and when the blocks for the component appear, drag the **whenRecordBtn.Click** block ❶ to the Viewer. Then, in the Blocks pane, click **Camcorder1**, drag another **callCamcorder1.RecordVideo** block ❷ to the Viewer, and snap it inside the **whenRecordBtn.Click** block next to the word do. These blocks call the Camcorder component's built-in RecordVideo method to open the phone's video camera when the RecordBtn is clicked, just as we planned.

Now live-test to see how these blocks work. When you click **Record**, your phone's video camera opens and you can then click the camera's record button to record a video. But don't record just yet! First we need to tell the app what to do after the camera records. If you close the video camera again without recording, you'll likely see the error saying you didn't record a video, which is okay for now.

STEP 3: CHOOSING THE VIDEO CLIP TO PLAY

Let's program step 3 of the app so that, after the user records a video, the app knows what to do with the recording. We want the app to set the *path* or storage location of the video as the source for the VideoPlayer component, which will ensure the recording will play next in the video player. Here are the blocks that handle this *after recording* event.

In the Blocks pane, click **Camcorder1**, and when the blocks for the component appear, drag the **whenCamcorder1.AfterRecording** block ❶ to the Viewer. Then, in the Blocks pane, click the **VideoPlayer1** component, drag the **setVideoPlayer1.Sourceto** block ❸ to the Viewer, and snap it inside the **whenCamcorder1.AfterRecording** block next to the word do.

The clip ❷ on the whenCamcorder1.AfterRecording block is an *event parameter*, which is a container that holds an *argument* or information provided about an event. This information can only be used in the event handler that provides it. The clip parameter holds the path to the video clip recorded by Camcorder1 in part 1 or 2 of the app, and the video player can't play the video until you set its *source*—the path to the file for it to play—to the path of the recorded clip.

To set the source, mouse over the **clip** parameter until you see a get clip block ❹. Drag the **get clip** block to the Viewer, and snap it to the right side of the **setVideoPlayer1.Sourceto** block. The blocks in this step set the path to the last recorded video clip as the source for VideoPlayer1.

STEP 4: PLAYING THE VIDEO CLIP

Finally, let's program step 4 of the app, telling it what to do when the user clicks WatchBtn. When the user clicks this button, we want the app's video player to start playing the user's recorded video. Here is the code for programming this event handler.

In the Blocks pane, click WatchBtn, and when the blocks for the component appear, drag the **whenWatchBtn.Click** block ❶ to the Viewer. Then, in the Blocks pane, click VideoPlayer1, drag the **callVideoPlayer1.Start** block ❷ to the screen, and snap it inside the **whenWatchBtn.Click** block next to the word do. These blocks tell the app to call the built-in Start method for VideoPlayer1 to start the video player when the watch button is clicked, as planned.

Now live-test your completed app! Open the app on your phone, and you should see the splash screen for 5 seconds before the record and watch buttons and VideoPlayer1 appear and your phone's video camera opens. Take a video and then click **Watch** to view the video in the app's integrated video player. If you placed your blocks as shown in the code examples, you should be able to record and watch videos over and over again, and you've successfully created the "Practice Makes Perfect" app!

SUMMARY

In this chapter, you built the "Practice Makes Perfect" app, where users can record, watch, and re-record temporary video clips of themselves practicing speeches or songs. In the process, you examined how App Inventor uses event-driven programming to set app behavior, and you became familiar with different types of events and event handlers. You also learned about App Inventor's Boolean value blocks and event parameters and used both to set values in this app.

In the next chapter, you'll learn how to animate images; use random numbers; and create, set, and change the values of *variables* using App Inventor's Animation, Math, and Variables blocks. We'll use these tools to create a noisy, animated "Fruit Loot" game app, where players get points for catching fruit as it drops down the screen.

ON YOUR OWN

Save new versions of "Practice Makes Perfect" as you modify and extend it working on these exercises. You can find solutions online at *https://nostarch .com/programwithappinventor/*.

1. Extend the app so that the user can record, watch, and compare two practice video clips side by side. Which Layout and other component(s) would you need to drag to the Viewer to enable this? How will your blocks change?

2. Change the app so that it records and plays sound clips instead of videos. What components and blocks would you use to record and play sound?

3

FRUIT LOOT: CREATING A SIMPLE ANIMATED GAME

In this chapter, you'll create a simple game called "Fruit Loot" that uses components from the Drawing and Animation, Sensors, and Media drawers to let players catch falling fruit.

You'll program these components with App Inventor's built-in Math and Variables blocks and component-specific blocks so that the game will use *animation*, or movement, with corresponding sound effects; unpredictability to make the game challenging; and the ability to keep score so players can see how well they're doing.

First, let's explore the key components and underlying programming concepts that you'll use to create the game.

ANIMATING AND MOVING RANDOMLY

To play the game, a player moves a fruit picker character back and forth across the screen trying to catch pieces of falling fruit. The pieces of fruit fall continuously at random speeds from random points at the top of the screen. Because of this random animation, players won't know exactly where to move the picker to catch the fruit. This unpredictability should challenge players and keep them engaged.

PROGRAMMING MOVING IMAGES

We'll use the Canvas and ImageSprite components from the Drawing and Animation drawer to create the moveable game character and constantly dropping fruit. The Canvas component is a layer or sheet that we place on the app's screen so users can draw. The Canvas is also where *sprites*, which are flat images, can move around. The game character and different pieces of fruit are all ImageSprites, which we'll place on a Canvas to make them move, collide with other sprites, and bounce off the edge of the screen.

The height and width of the Canvas are measured in *pixels*, a unit of measurement used in computer graphics, and App Inventor uses a common computer screen coordinate system to determine the exact location of an ImageSprite on the Canvas. In that coordinate system, the top-left point of the ImageSprite is located at the point represented by its x- and y-coordinates or properties (X,Y). The X property is the image's distance in pixels from the Canvas's left edge, and the Y property is the picture's distance from the Canvas's top edge.

NOTE *In math class, you may have plotted points on a coordinate plane that contains four quadrants. App Inventor's coordinate system is like the lower-right quadrant in that coordinate plane, where the point of origin (0, 0) is at the top left, and the size of the x-coordinate increases from left to right, while the size of the y-coordinate increases from top to bottom. The difference is that, in the math plane, the increasing y-coordinate numbers are negative, while they're positive in App Inventor.*

As shown in Figure 3-1, in App Inventor's coordinate system, an Image Sprite's X property value increases as the graphic moves to the right, and its Y property value increases as it moves down.

When adding an ImageSprite to the Canvas, we set its initial X and Y property values to the point where we place it or to other values we enter into the Designer window's Property pane. To move the ImageSprite, we use program blocks to change either property value.

For this game, you'll program button click event handlers to let players move the fruit picker. Also, to constantly animate the fruit, you'll program the Clock so the fruit moves automatically at a time interval you'll set. JavaScript and other programming languages handle animation the same way, by having images change location in response to user or automated actions.

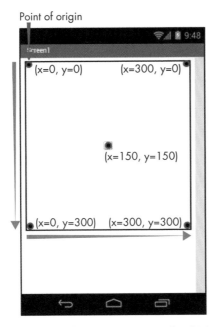

Point of origin

(x=0, y=0)   (x=300, y=0)

(x=150, y=150)

(x=0, y=300)   (x=300, y=300)

Figure 3-1: The Viewer screen with a 300×300 pixel Canvas showing the point of origin and X and Y property values of different points

SETTING UP RANDOM APPEARANCE, SPEED, AND LOCATION

Like other games that deal with chance, this game relies on randomness to keep players from plotting exactly how, when, or where to make their next moves. Because of the need for randomness in games and other applications, such as simulations, most traditional programming languages include *pseudorandom number generators*, which are functions based on mathematical algorithms.

In its built-in Math blocks, App Inventor provides two pseudorandom number generators. The blocks also include *arithmetic operators* that perform multiplication, division, addition, and subtraction functions on *operands* or values, just like similar operators in other programming languages. In your game, you'll combine one of App Inventor's pseudorandom number generators, called the *random integer block*, with arithmetic operator blocks to develop simple formulas to ensure that the appearance, speed, and location of each piece of falling fruit will be determined randomly.

Thanks to these formulas, although our players will quickly realize that fruit drops constantly, they won't know where and at what speed it will fall, keeping them from scoring points too easily.

DECLARING AND INITIALIZING VARIABLES

As players earn points, we'll need a way to let them know their score. We'll do this by declaring and initializing a couple of *variables*—uniquely named containers of memory that programmers create to hold values that can change, or vary. Variables allow us to store necessary information that we can update from time to time as conditions in the app change. We can use the unique variable name to refer to that changing information throughout our code and perform operations on the information as the app runs, no matter what value the variable holds at any given time.

For instance, in your game, you'll store the score in a variable and compute and display the changing score during the game. To compute the score, you'll use a Math addition operator block to *increment* the score, or increase it by a fixed number—in this case, 1—whenever a player earns a point.

In traditional programming languages, you must follow specific syntactical rules to *declare*, or create, a variable and to *initialize* it, or assign its first value. In some languages, you also have to identify the type of data the variable will hold when you create it. In App Inventor, you must declare and initialize variables using the built-in Variables blocks, and you can store *strings* (sets of characters that can include letters, numbers, and other characters), individual numbers, Boolean values, and lists by snapping in blocks from the Text, Math, Logic, and Lists drawers.

As you work with variables, you'll notice that they're a lot like component properties in that both variables and properties hold data that can be set, reset, and accessed by the blocks used in an app. In fact, as soon as you create a variable, App Inventor creates getter and setter blocks for it, similar to those available for properties, and adds them to the Variables blocks drawer.

In your game, you'll create variables that have a *global scope*, which means you can use them in all of your event handlers. In later chapters, you'll experiment with *local* variables, which you'll create within an event handler or procedure for use only within that handler or procedure. All programming languages use global and local variables.

BUILDING THE "FRUIT LOOT" APP

Now that you understand how to create variables and program animation and randomness in App Inventor, you're ready to create "Fruit Loot."

To get started, log into App Inventor following the instructions outlined in "Getting Started with App Inventor" on page xviii. In the dialog for the project name, enter `FruitLoot` without any spaces, and then click **OK**.

DECOMPOSING "FRUIT LOOT"

In "Fruit Loot," the player moves a fruit picker across the screen to catch rapidly and randomly dropping fruit. The player earns a point for each fruit caught and sees the score on the screen. We can decompose the game activity into five steps:

1. When a player presses the start button, start the game.
2. When the Clock timer fires, drop fruit from the top of the Canvas at different speeds.
3. When a fruit hits the bottom of the Canvas, return it to a random point at the top of the Canvas and display another fruit at random. Increase the total fruits dropped by one.
4. When a player clicks the left and right buttons, move the picker left and right to catch the falling fruit.
5. When the picker catches a fruit, play a sound, increase the player's score by one, display the score, and hide the fruit.

   Here are the components you'll need:

- **Button** (3) for the player to click to manually start the action and play the game
- **Canvas** to enable use of ImageSprites and game animation
- **Clock** to fire after the player clicks the start button and move ImageSprites at a set interval
- **HorizontalArrangement** (2) to hold start button, score label, and play buttons
- **ImageSprite** (4) to display moving images
- **Label** to display Variable values
- **Sound** to play the game sound effect
- Variable (2) to store game data

LAYING OUT "FRUIT LOOT" IN THE DESIGNER

Now let's lay out the app in the Designer. First, change the Screen's horizontal alignment so that everything we place on it will be centered. Click **Screen1** in the Components pane, click the drop-down arrow under **AlignHorizontal** in the Properties pane, and select **Center: 3**.

Next, let's add a background image to the Screen by clicking the text box under **BackgroundImage** in the Properties pane. Follow the image upload instructions outlined in "Uploading a Picture" on page 27 to upload *fence -tree.png*, which comes with the resources for this book. You can download the resources from *https://nostarch.com/programwithappinventor/*.

Now let's change the screen *orientation*, which generally means whether the screen displays vertically (in *portrait* mode) or horizontally (in *landscape* mode). By default, ScreenOrientation is set to Unspecified, which means that the orientation changes depending on how a user rotates the device.

To give our picker ImageSprite a wider screen area to move across to catch fruit, let's change the orientation to landscape mode to make sure the screen displays horizontally regardless of how the device is held. Click the drop-down arrow under **ScreenOrientation** and select **Landscape**. Also, unclick the checkbox under both **ShowStatusBar** and **TitleVisible** to keep the device status bar and Screen title from showing and taking up space when the game displays on a device.

ADDING AND ARRANGING USER INTERFACE COMPONENTS

Since we have limited vertical screen space in landscape orientation mode, we need to take up as little of that space as possible with our user interface components. But we still need to make sure those components are easy for players to see and use. To accomplish this, let's place our Buttons and Label in two HorizontalArrangements, one across the top of the screen and one across the bottom.

Drag two **HorizontalArrangement**s from the Layout drawer onto the Viewer. Then, click each in the Components pane, and rename the first one **TopArrangement** and the second **BottomArrangement**. Then, in the Properties pane, center both of their horizontal alignments by clicking the drop-down arrow under **AlignHorizontal** and selecting **Center: 3**, which should center all the components we place inside. Next, make **BottomArrangement**'s width **Fill parent**, the same way you did with components in Chapters 1 and 2, so that it stretches all the way across the screen.

Now drag a **Button** and a **Label** from the User Interface drawer into **TopArrangement**. Then, in the Components pane, click **Button1** and rename it **StartBtn**, and in the Properties pane, change its text size to 18 point by clicking the text box under **FontSize**, deleting the current number, and entering **18**. Also change the default text showing on StartBtn by clicking the text box under **Text**, deleting the current text, and entering **Start the Fruit Loot Game**. Then, in the Components pane, click **Label1**, and in the Properties pane, remove the Label's default text by clicking the text box under **Text** and deleting the current text so no text will show until the game starts. Then, center the text by clicking the drop-down arrow under **TextAlignment** and selecting **Center: 1**.

Next, drag the remaining two **Button**s from the User Interface drawer into **BottomArrangement**, click each in the Components pane, and rename the one on the left **LeftBtn** and the other **RightBtn**. Also make the width for each Button **Fill parent**, which makes each take up half the width of Bottom Arrangement. Now change the text showing on LeftBtn to **<<<< Left** and RightBtn to **Right >>>>**.

Finally, make the background color orange for all three Buttons and the Label by clicking the box under **BackgroundColor** and selecting **Orange** from the color list dialog. Also make the text displaying on the Buttons and Label bold by clicking the checkbox under **FontBold**. Next, update the font size on all but StartBtn by clicking the text box under **FontSize** and entering **10** to replace the existing number. Finally, for **BottomArrangement**, unclick the checkbox under **Visible** so LeftBtn and RightBtn won't show when the app opens.

SETTING UP THE CANVAS AND IMAGESPRITES

Now, click the **Drawing and Animation** drawer and drag a Canvas onto the Viewer between TopArrangement and BottomArrangement. Remember that you must place a Canvas on the Screen before you can add any other Drawing and Animation component. In the Properties pane, make the Canvas transparent so it doesn't hide the background image by clicking the box under **BackgroundColor** and then clicking **None** when the color list dialog opens. Then make its height and width **Fill parent**.

Now drag four ImageSprites from the Drawing and Animation drawer onto the Canvas, click the ImageSprites in the Components pane, and rename the first three **FruitSprite1**, **FruitSprite2**, and **FruitSprite3** and the last **Picker Sprite**. Next, under **Picture** in the Properties pane, for the fruit ImageSprites, upload *1.png*, *2.png*, and *3.png*, and for the picker ImageSprite, upload *picker .png*. (All of these images come with the resources for this book.)

Finally, you can either drag the ImageSprites around the Canvas or enter numbers in the text boxes under their X and Y properties to position them on the Canvas the way they should appear when the game starts. We want the fruit ImageSprites spread out evenly across the top of the Canvas and Picker Sprite in the center at the bottom. To place the components this way on a screen that's approximately 450 pixels wide, enter the numbers in Table 3-1 into the Property pane text boxes under **X** and **Y** for each ImageSprite.

Table 3-1: Initial X and Y Property Values for "Fruit Loot" ImageSprites on a 450-pixel-wide screen

ImageSprite	X property	Y property
FruitSprite1	10	0
FruitSprite2	230	0
FruitSprite3	440	0
PickerSprite	180	150

Now you're ready to add and adjust the non-visible components.

ADDING AND PREPARING NON-VISIBLE COMPONENTS

From the Media drawer, drag in a Sound component, and from the Sensors drawer, drag in a Clock component. In the Components pane, click the Sound component, and in the Properties pane, set the media clip that it will play by clicking the text box under **Source** and uploading the *clunk.mp3* file that comes with the book resources.

Then click the **Clock**, replace its default TimerInterval property by entering 150, and unclick the checkbox under **TimerEnabled** so the timer won't start when the app opens. Shortly, we'll program the blocks to enable the timer once the player clicks StartBtn.

At this point, your screen should look like Figure 3-2.

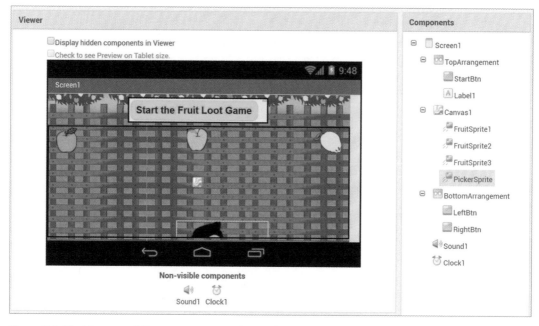

Figure 3-2: The Viewer and Component panes after we lay out "Fruit Loot"

Once your screen looks like it should, click the **Canvas** in the Components pane, and unclick the checkbox under **Visible**. This way, none of the ImageSprites should show when the app opens. Next, we'll program the blocks to make these components, LeftBtn and RightBtn, visible once the player clicks StartBtn.

PROGRAMMING "FRUIT LOOT" IN THE BLOCKS EDITOR

Now that you've laid out all the components, you can move to the Blocks Editor to program the app. For "Fruit Loot" we'll program 10 event handlers. Three respond to events generated by the user's button clicks. One directs the app's action after a timer goes off at the time interval we've set. The rest respond to ImageSprites reaching the edge of the Canvas or colliding with one another.

You'll notice that most of the event handlers contain duplicate code. We're programming them this way because you haven't yet learned the advanced programming structures that would eliminate the repetition.

As you learn about those structures in later chapters, we'll be able to revisit the "Fruit Loot" code and *refactor* it, which means to restructure and improve it.

Click the **Blocks** button to switch to the Blocks Editor, and let's begin programming the five steps of "Fruit Loot" in order.

STEP 1: STARTING THE GAME

We start by telling the app what to do when the player clicks StartBtn. That's when we want the StartBtn to disappear, the Canvas with its ImageSprites and BottomArrangement with its Buttons to appear, and the Clock's timer to begin to fire.

Here is the button click event handler with its four setter blocks that do what we want in step 1.

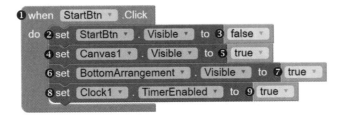

In the Blocks pane, click **StartBtn** and, when the blocks for the component appear, drag the **whenStartBtn.Click** event handler block ❶ to the Viewer. Then, in the Blocks pane, click **StartBtn** again, and drag its **set StartBtn.Visibleto** block ❷ into the **whenStartBtn.Click** block next to the word do. Next, in the Blocks pane, click the **Logic** blocks drawer, drag the **false** block ❸ to the Viewer, and snap it to the right side of the **setStartBtn .Visibleto** block. These blocks set StartBtn's Visible property to false so that it disappears after the player clicks the start button.

Next, click **Canvas1**, drag the **setCanvas1.Visibleto** block ❹ to the Viewer, and snap it inside the **whenStartBtn.Click** block under the setStartBtn.Visibleto block. Then, in the Blocks pane, click the **Logic** blocks drawer again, drag the **true** block ❺ to the Viewer, and snap it to the right side of the **setCanvas1 .Visibleto** block. These blocks set the Visible property for Canvas1 and its contents to true so the ImageSprites will appear after the player clicks the start button.

Then, click **BottomArrangement** in the Blocks pane, drag the **setBottom Arrangement.Visibleto** block ❻ to the Viewer, and snap it inside the **whenStart Btn.Click** block under the setCanvas1.Visibleto block. Then click the **Logic** blocks drawer again, drag another **true** block ❼ to the Viewer, and snap it to the right side of the **setBottomArrangement.Visibleto** block. These blocks set the Visible property for BottomArrangement to true, making the buttons inside of it appear after the player clicks the start button.

Finally, click **Clock1**, drag the **setClock1.TimerEnabledto** block ❽ to the Viewer, and snap it inside the **whenStartBtn.Click** block under the setBottom Arrangement.Visibleto block. Then, in the Blocks pane, drag another **true** block ❾ from the **Logic** blocks drawer, and snap it to the right side of the `setClock1.TimerEnabledto` block. These blocks set the value of `Clock1`'s `Timer Enabled` property to true. This starts `Clock1`'s timer, which will move the fruit `ImageSprites` down the `Canvas` the entire time the game is in play.

Together, the blocks for step 1 start the game. In sum, when the player clicks the start button, the blocks set `StartBtn`'s `Visible` property to false, set the `Visible` properties of the `Canvas` with `ImageSprites` and the `Horizontal Arrangement` with play buttons to true, and set the `Clock`'s `TimerEnabled` property to true.

To see how these blocks work, live-test with a device, as outlined in "Live-Testing Your Apps" on page xxii. Once you click **Connect ▶ AI Companion** in the top menu bar and scan the QR code with your device's Companion app, your "Fruit Loot" game should open on your device. As long as your blocks are placed as shown in the code examples, you should see the start button until you click it, when it disappears as the other game components appear. For now, nothing else should happen. Leave the game open on your device to keep live-testing.

STEP 2: MAKING FRUIT DROP AT RANDOM

Now let's program step 2 of the app and tell it what to do each time the `Clock`'s timer fires. This is when we want the fruit to drop at varying speeds every 150 milliseconds—the `TimerInterval` we set in the Property pane in the Designer.

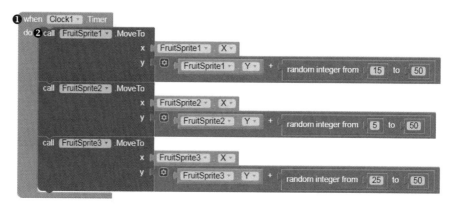

In the Blocks pane, click **Clock1** and, when the blocks for the component appear, drag the **whenClock1.Timer** block ❶ to the Viewer. Then, in the Blocks pane, click **FruitSprite1**, drag the **callFruitSprite1.MoveTo** method block ❷ to the Viewer, and snap it inside the **whenClock1.Timer** block next to the word do.

NOTE   *In both the Components pane in the Designer and the Blocks pane in the Blocks Editor, if you don't see a component that you've nested within a parent component— for example, an ImageSprite placed on a Canvas or a Button dragged within a HorizontalArrangement—you'll find it by clicking the plus sign to the left of the parent.*

Setting X and Y Values for FruitSprite1

Let's look closer at the callFruitSprite1.MoveTo method block we've placed within the Clock1 Timer event handler.

You'll notice that the block requires us to insert values for its x and y *method parameters*, which are pieces of information the method must have to operate. This means the ImageSprite's MoveTo method cannot move Fruit Sprite1 until we supply *arguments*, or values, for the x- and y-coordinates of the point where we want the ImageSprite to move.

For our "Fruit Loot" game, we want the fruit ImageSprites to move down only, meaning we'll change their y-coordinates but not their x-coordinates. To keep the same X value, click **FruitSprite1**, drag its **FruitSprite1.X** getter block ❶ to the Viewer, and snap it into the method block's **x** socket. This tells the app that, when it moves FruitSprite1, it should get the current X value for FruitSprite1 and keep that X value the same.

To provide the argument for the y-coordinate of the point where we want FruitSprite1 to move, click the **Math** blocks drawer, drag out an addition operator block ❷, and snap it into the method block's **y** socket. Then, click **Fruit Sprite1** and drag its **FruitSprite1.Y** getter block ❸ into the addition block's left socket, and click the **Math** blocks drawer and drag a random integer block ❹ into the addition block's right socket. This tells the app that, when it moves FruitSprite1, it should increase the current value of its y-coordinate by a random number of pixels to move the ImageSprite down the Canvas.

Dropping FruitSprite1 at Random Speeds

The random integer block generates the random number of pixels—from between the specified range of 15 to 50—that we want FruitSprite1 to fall.

To set that range of numbers in the random integer block, click the default **1** in its left socket and replace it by entering **15**, and click the default **100** in its right socket and replace it by entering **50**.

Now, altogether, our callFruitSprite1.MoveTo method block with the x and y parameters we've set tells the app that, when it moves FruitSprite1, we want the ImageSprite's X value to stay the same and its Y value to move from its current y-coordinate down a random number of pixels between 15 and 50. This randomness ensures that the ImageSprite's speed will be unpredictable, because, when the Clock's timer fires every 150 milliseconds, FruitSprite1 will travel at a speed anywhere from a slower 15 pixels per 150 milliseconds (100 pixels per second) to a faster 50 pixels per 150 milliseconds (333 pixels per second).

Copying Blocks for FruitSprite2 and FruitSprite3

To complete the blocks for step 2, we now can copy the callFruitSprite1
.MoveTo block and adjust it for FruitSprite2 and FruitSprite3. Right-click the
callFruitSprite1.MoveTo method block to duplicate it for **FruitSprite2**, and
snap the duplicate in under the original.

In the duplicate blocks, use the drop-down arrows ❶ in the **callFruit
Sprite1.MoveTo, FruitSprite1.X**, and **FruitSprite1.Y** blocks to change to **Fruit
Sprite2**. Also change the number in the left random integer block socket to **5**.
These blocks now program FruitSprite2 to move down some unknown
number of pixels between 5 and 50 when the Clock's timer fires every
150 milliseconds.

Next, right-click the **callFruitSprite1.MoveTo** block, make another copy
to use for FruitSprite3, and snap the duplicate in under the **callFruitSprite2
.MoveTo** block.

In the duplicate, use the drop-down arrows ❷ in the **callFruitSprite1
.MoveTo, FruitSprite1.X**, and **FruitSprite1.Y** blocks to change to **FruitSprite3**,
and change the number in the random integer block's left socket to **25**. These
blocks program FruitSprite3 to move down a random number of pixels
between 25 and 50 every 150 milliseconds.

Now the blocks for step 2 should move the three fruit ImageSprites down
the Canvas every 150 milliseconds at random speeds.

Live-test to see how these blocks work. When you click StartBtn, you
should see the three fruit ImageSprites drop to the bottom of the screen,
where they stay and all movement stops. If any ImageSprite fails to move, you
need to debug. In this instance, you may not have changed your references
to the correct ImageSprite when you duplicated the MoveTo blocks. Make any
necessary corrections, and test again. Once step 2 is working, move to the
next step, where we'll tell the game what to do when the fruit ImageSprites
reach the bottom of the Canvas.

STEP 3: CREATING MORE FALLING FRUIT AND COUNTING DROPPED FRUIT

Let's now program step 3 of the app. In this part, when a fruit ImageSprite
reaches the bottom edge of the Canvas, we want the app to move the

ImageSprite back up to a random point along the very top of the Canvas, have the ImageSprite display a random picture of fruit, and add 1 to the total number of times an ImageSprite hits the edge.

We'll use a global variable to store and update that total number, and we'll start our code for this step by creating and initializing that variable.

initialize global `fruitsDropped` to `0`

Click the **Variables** blocks drawer and drag an `initialize global name` block to the Viewer. Click `name` and replace it with the name of our variable, `fruitsDropped`. Then drag a `0` number block from the **Math** drawer and snap it to the right side of the `initialize global fruitsDropped` block. This declares and initializes the variable you'll use to store and update the total number of pieces of fruit dropped in your game. Because the variable is global and can be used by all your event handlers, it stands alone in the code, outside of all your event handler blocks.

Now let's program the event handler for this step. Here are the blocks that handle this EdgeReached event for FruitSprite1.

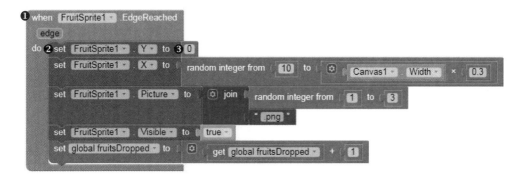

In the Blocks pane, click **FruitSprite1** and, when the blocks for the component appear, drag the **whenFruitSprite1.EdgeReached** block ❶ to the Viewer. Then, in the Blocks pane, click **FruitSprite1** again, drag the **setFruitSprite1.Y** block ❷ to the Viewer, and snap it inside the **whenFruitSprite1.EdgeReached** block next to the word do.

Then, click the **Math** blocks drawer, drag a **0** number block ❸ to the Viewer, and snap it to the right of the **setFruitSprite1.Y** block. So far, once FruitSprite1 reaches the edge of the Canvas, these blocks move FruitSprite1 right back up to y-coordinate 0, which is the very top of the Canvas.

Moving Fallen Fruit Back Up to a Random Place

We now need to make sure the code also moves FruitSprite1 to an unpredictable x-coordinate using setter blocks, which keeps your game interesting.

❶ set `FruitSprite1` . `X` to ❷ random integer from `10` to ❸ ❹ `Canvas1` . `Width` × ❺ `0.3`

In the Blocks pane, click the **FruitSprite1** component, drag the **setFruit Sprite1.X** block ❶ to the Viewer, and snap it inside the **whenFruitSprite1.Edge Reached** block under the setFruitSprite1.Y block. Then, click the **Math** blocks drawer, snap a random integer block ❷ to the right of the **setFruitSprite1.X** block, and click the default **1** in the random integer block's left socket and replace it by entering **10**. Then, delete the default **100** in the random integer block's right socket, and replace it with a multiplication operator block ❸, also from the Math drawer. Next, in the Blocks pane, click the **Canvas1** component, drag the **Canvas1.Width** getter block ❹ to the Viewer, and snap it into the multiplication block's left socket; then, drag a **0** number block ❺ from the **Math** drawer to the Viewer, click its default **0**, replace it by entering **0.3**, and then snap the **0.3** number block into the multiplication block's right socket.

These blocks set the X value for FruitSprite1 once it reaches the edge of the Canvas. To avoid collisions with other fruit ImageSprites, we want this first fruit ImageSprite to drop somewhere in the left third of the Canvas only. These blocks ensure that by setting the new X position to a random number of pixels between 10 and the width of the Canvas multiplied by 0.3, which is a little less than one-third of the Canvas width. For instance, if the width of the Canvas is 450 pixels, the new X position will be anywhere between 10 and (450 × 0.3) pixels, which equals 135 pixels.

Dropping Random Fruit Images

Next, to keep your game unpredictable, you need to make sure the code randomly changes the type of fruit dropped after FruitSprite1 moves back up to the top of the Canvas. To do this, you'll use setter blocks that set the Picture property for FruitSprite1 to a random image.

Click the **FruitSprite1** component in the Blocks pane, drag the **setFruit Sprite1.Picture** block ❶ to the Viewer, and snap it inside the **whenFruitSprite1 .EdgeReached** block under the setFruitSprite1.X block. Then, click the **Text** blocks drawer, drag a join block ❷ to the Viewer, and snap it to the right of the **setFruitSprite1.Picture** block, which will allow you to set the name for the picture by joining two strings.

For the join block's top input, drag in another random integer block ❸ from the Math blocks drawer, click the default **100** in its right socket, and replace it by entering **3**. For the join block's second input, drag in an empty string block ❹, the first block in the Text blocks drawer. Then click the string block's text area and enter **.png**.

These blocks set the name of the image to use as the picture source for FruitSprite1 after it reaches the Canvas edge. Since we've named the three uploaded fruit images *1.png*, *2.png*, and *3.png*, we can use the random integer block to generate the number 1, 2, or 3 that is part of the image name. This should make the app continually display a randomly selected image on Fruit Sprite1 each time it drops from the top of the Canvas.

Making Sure Fruit is Visible

We also need to make sure FruitSprite1 and the other fruit ImageSprites are visible once they move back up to the top of the Canvas, because later we'll make them invisible if they hit the picker ImageSprite. Here is the setter block that turns the Visible property on.

```
1 set FruitSprite1 . Visible . to 2 true .
```

Click **FruitSprite1**, drag the **setFruitSprite1.Visibleto** block ❶ to the Viewer, and snap it inside the **whenFruitSprite1.EdgeReached** block under the setFruitSprite1.Pictureto block. Then, in the Blocks pane, click the **Logic** blocks drawer, drag the **true** block ❷ to the Viewer, and snap it to the right side of the **setFruitSprite1.Visibleto** block. These blocks reset the ImageSprite's Visible property to true in case it collides with the picker ImageSprite, after which our blocks in step 5 will set it to false.

Counting the Number of Fruits Dropped

Finally, we need to program the app to keep track of how many fruits are dropped. Each time a fruit ImageSprite hits the bottom of the Canvas, the game should add 1 to the value of fruitsDropped, which is the variable that keeps track of the number of times an ImageSprite hits the edge. The following blocks increment the value of fruitsDropped.

```
1 set global fruitsDropped . to 2 ⚙ 3 get global fruitsDropped . + 4 1
```

Mouse over the **initialize global fruitsDropped** block that we placed at the beginning of this step, drag the **set global fruitsDropped** block ❶ to the Viewer, and snap it inside the **whenFruitSprite1.EdgeReached** block under the setFruitSprite1.Visibleto block. Then drag an addition operator block ❷ from the Math drawer and snap it to the right of the **set global fruitsDropped** block. Fill that addition block's sockets by mousing over the **initialize global fruitsDropped** block, dragging the **get global fruitsDropped** block ❸ into the addition block's left socket, and dragging a **1** number block ❹ from the Math drawer into the addition block's right socket. These blocks keep track of the game's total number of fruits dropped by adding 1 to the current value of the fruitsDropped variable each time FruitSprite1 reaches the edge of the Canvas.

Copying Blocks for FruitSprite2 and FruitSprite3

Our final task for step 3 is to duplicate our code to program similar EdgeReached event handlers for FruitSprite2 and FruitSprite3.

For FruitSprite2, right-click the **whenFruitSprite1.EdgeReached** block and select **duplicate**. When you make this duplicate, you'll see a red X appear to the left of the word when in both the original and duplicate event handlers. This red X warns you that you have two event handlers for the same event, which is not allowed. Once you change the duplicate handler's event, the red X should disappear. To change the event, use the drop-down arrow in every block where you see FruitSprite1 and change to **FruitSprite2**.

The only other adjustment we need to make is to set an X property for FruitSprite2 that avoids collisions with the other fruit ImageSprites when it moves down the Canvas. To accomplish this, make sure this second fruit ImageSprite consistently drops somewhere in the middle third of the Canvas by changing the FruitSprite2 X value.

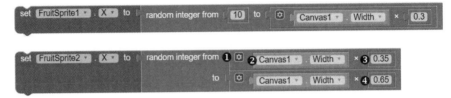

NOTE *App Inventor provides a way for you to display the* random integer, join, *and other blocks that require multiple inputs with either* inline inputs, *as shown in the* random integer *block inside the* FruitSprite1 *setter, or* external inputs, *as shown in the* random integer *block inside the figure's* FruitSprite2 *setter, which takes up less horizontal space. You can right-click a block to switch between inline and external inputs.*

Replace the **10** number block in the **random integer** block's top socket with a multiplication operator block ❶ from the Math drawer. Then fill the multiplication block's sockets by clicking the **Canvas1** component in the Blocks pane, dragging the **Canvas1.Width** block ❷ into its left socket, and dragging a **0.35** number block ❸ from the Math drawer into its right socket. Then, in the random integer block's bottom socket, change the **0.3** number block to a **0.65** number block ❹.

These blocks set the new X position for FruitSprite2 to a random number of pixels between the width of the Canvas multiplied by 0.35 and the width of the Canvas multiplied by 0.65, which is some random point in the middle third of the Canvas. For instance, if the width of the Canvas is 450 pixels, the new X position will be anywhere between 450 × 0.35 pixels, which equals 158 pixels, and 450 × 0.65 pixels, which equals 293 pixels.

Now, to create the EdgeReached event handler for FruitSprite3, right-click the **whenFruitSprite2.EdgeReached** block and select **duplicate**. In the duplicate blocks, be sure to use the drop-down arrow in every block where you see FruitSprite2 and change to **FruitSprite3**.

To avoid collisions with the other fruit ImageSprites, we'll also need to change the FruitSprite3 X value so this third fruit ImageSprite consistently drops somewhere in the right third of the Canvas.

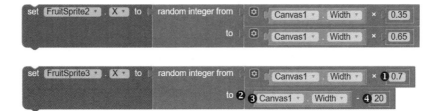

To do this, replace the **0.35** number block in the right socket of the first multiplication block with a **0.7** number block ❶. Also, replace the second multiplication block with a subtraction operator block ❷ from the Math blocks drawer, and fill the subtraction block by dragging the `Canvas1.Width` block ❸ into its left socket and a **20** number block ❹ into its right socket.

These blocks set the new X position of `FruitSprite3` to a random number of pixels between the width of the `Canvas` multiplied by 0.7 and the width of the `Canvas` minus 20 pixels, which is some random point in the right third of the `Canvas`. For instance, if the width of the `Canvas` is 450 pixels, the new X position will be anywhere between 450 × 0.7 pixels, which equals 315 pixels, and 450 – 20 pixels, which equals 430 pixels.

Altogether, the blocks for step 3 move each fruit `ImageSprite` to a random point at the very top of the `Canvas`, have that `ImageSprite` display a random picture of fruit, and increase the count of total fruits dropped by 1 each time an `ImageSprite` hits the bottom of the `Canvas`, just as we planned.

Now live-test the game again. This time, when you click `StartBtn`, you should see the three fruit `ImageSprites` drop to the bottom of the screen continuously, randomly changing the image displayed. You'll also notice that the speed at which each `ImageSprite` drops and its X location changes with each drop.

If any of the `ImageSprites` fail to move, or if two or more appear to drop in the same third of the `Canvas`, debug your code. Here again, you may not have made the correct changes to your duplicate blocks. Make any necessary corrections, and test again. Once step 3 is working, let's move to the next part, where we'll program `PickerSprite`'s movement.

STEP 4: LETTING PLAYERS MOVE THE PICKER TO CATCH THE FRUIT

Now let's program step 4 of the app, telling it what to do when the player clicks the left and right play buttons. When the player clicks `LeftBtn`, we want `PickerSprite` to move to the left 50 pixels, and when the player clicks `RightBtn`, we want `PickerSprite` to move 50 pixels to the right.

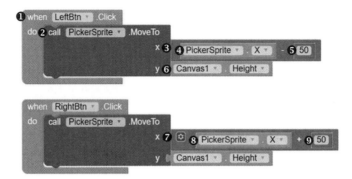

In the Blocks pane, click **LeftBtn** and, when the blocks for the component appear, drag the `whenLeftBtn.Click` block ❶ to the Viewer. Then, in the Blocks pane, click **PickerSprite**, drag the `callPickerSprite.MoveTo` block ❷ to the Viewer, and snap it inside the `whenLeftBtn.Click` block next to the word do.

Now we need to provide the MoveTo block's method parameters to tell the app where we want to move PickerSprite, keeping in mind that, for this game, we want PickerSprite to move from side to side only, along the very bottom of the Canvas. That means we want to change its x-coordinate but leave its y-coordinate at the bottom. To do this, click the **Math** blocks drawer, drag a subtraction operator block ❸ to the Viewer, and snap it to the right of **x**. Then, click **PickerSprite** and drag its **PickerSprite.X** block ❹ into the subtraction block's left socket, and drag a **50** number block ❺ from the Math drawer into the subtraction block's right socket.

These blocks tell our app to move PickerSprite's x-coordinate left to its current location minus 50 pixels whenever the Button is clicked. For instance, if PickerSprite's x-coordinate is at 240 pixels, when the player clicks LeftBtn, the x-coordinate should move 50 pixels to the left to 190 pixels, since 240 pixels – 50 pixels = 190 pixels.

Next, click **Canvas1**, drag its **Canvas.Height** block ❻ to the viewer, and snap it into the **callPickerSprite.MoveTo** block's **y** socket. This tells the app that, when it moves PickerSprite, we want the ImageSprite's Y value to stay the value that equals the height of the Canvas, positioned at the bottom. For instance, if the Canvas is 300 pixels in height, these blocks will keep PickerSprite's y-coordinate at the Canvas's bottommost point, 300 pixels, when LeftBtn is clicked.

Now copy the **LeftBtn** event handler and modify the duplicate blocks to program **RightBtn**. First, in the duplicate event handler, be sure to use the drop-down arrow to change **LeftBtn** to **RightBtn**. Then replace the subtraction block after the letter x with an addition block ❼ from the Math drawer, click **PickerSprite** and drag its **PickerSprite.X** block ❽ into the addition block's left socket, and drag a **50** number block ❾ from the Math drawer into the addition block's right socket.

These blocks say move PickerSprite's x-coordinate to its current location plus 50 pixels when the button is clicked. So, if PickerSprite's x-coordinate is at 240 pixels when the player clicks RightBtn, the x-coordinate should move 50 pixels to the right to 290 pixels, since 240 pixels + 50 pixels = 290 pixels.

Now live-test the game again, and if LeftBtn and RightBtn don't work correctly after you click StartBtn, try debugging. LeftBtn and RightBtn should move PickerSprite back and forth across the screen, while fruit ImageSprites occasionally collide with PickerSprite. Since the player's goal in the game is to collide with, or "catch," the fruit, we need to program quite a bit of activity to take place when those fruit ImageSprites hit PickerSprite. We'll program that action in the next, and final, step.

STEP 5: HIDING CAUGHT FRUIT AND KEEPING SCORE

Now we'll program the last part of the game so that each time a fruit ImageSprite hits PickerSprite, the player "catches" the piece of fruit, hears a noise that sounds like the fruit hitting the picker's bucket, earns a point, and sees the total score displayed on the screen. We'll also hide the ImageSprite that hit PickerSprite so that, instead of continuing to fall to the bottom of the Canvas, the fruit looks like it landed in the picker's bucket.

To keep the player's score, we'll use a variable to store and update that information. Let's start our code for this step by creating and initializing the score variable.

initialize global `score` to `0`

Click the **Variables** block drawer and drag an `initialize global name` block to the Viewer. Click `name`, and replace it with the name of our variable, **score**. Then drag a **0** number block from the Math drawer and snap it to the right side of the `initialize global score` block. This declares and initializes the global variable we'll use to store and update the player's game score.

Playing a Sound When Fruit Hits the Picker

Let's now program the event handler for when a fruit ImageSprite hits PickerSprite.

In the Blocks pane, click **PickerSprite** and, when the blocks for the component appear, drag the `whenPickerSprite.CollidedWith` block ❶ to the Viewer. Then, in the Blocks pane, click **Sound1**, drag the `callSound1.Play` block ❷ to the Viewer, and snap it inside the `whenPickerSprite.CollidedWith` block next to the word do. This should play our clunking sound each time an Image Sprite hits PickerSprite.

Increasing and Displaying the Score

Next, let's place the blocks that increment and display the game score each time a fruit ImageSprite hits the picker.

Mouse over the **initialize global score** block, drag the **set global score to** block ❶ to the Viewer, and snap it inside the **whenPickerSprite.Collided With** block under the callSound1.Play block. Then drag an addition operator block ❷ from the Math drawer and snap it to the right of the **set global score to** block. Next, mouse over the **initialize global score** block again, and drag the **get global score** block ❸ into the addition block's left socket and a **1** number block ❹ from the Math drawer into its right socket. These blocks add 1 to the current value of the score variable each time a fruit ImageSprite collides with PickerSprite.

To display the score and also let the player know how many of the total number of dropped fruits PickerSprite has caught, click **Label1**, drag the **set Label1.Textto** block ❺ to the Viewer, and snap it inside the **whenPickerSprite .CollidedWith** block under the set global score to block. Then, click the **Text** blocks drawer, drag a **join** block ❻ to the Viewer, and snap it to the right of the **setLabel1.Textto** block.

Here, we'll join four strings to set the text and numbers we want Label1 to display, although by default the join block allows us to combine only two strings. Figure 3-3 shows how to change the block to create space for the additional inputs we'll need.

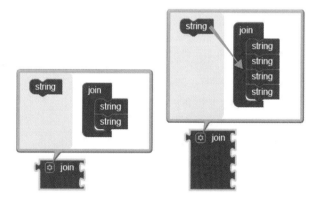

Figure 3-3: Adding inputs to the join *block*

Click the blue *mutator* icon to the left of the word join, and drag additional string blocks to the join block in the dialog that opens.

Now we can snap our four strings into the join block inputs. In the top input, drag in an empty string block ❼ from the Text blocks drawer and enter

`Score: `, making sure to include the space after the colon so that, when the strings combine, the characters won't run together without proper spacing. Then mouse over the **initialize global score** block and drag the **get global score** block ❽ into the join block's second input. In the join block's third input, drag in another empty string block ❾ and enter **out of**, leaving a space before out and after of. Then mouse over the **initialize global fruitsDropped** block and drag the **get global fruitsDropped** block ❿ into the join block's last input. These blocks display the number of points and total number of fruits dropped on Label1 for the player to see at the top of the screen. For instance, if the player's score is 6 points and a total of 20 fruits have dropped, the label should display "Score: 6 out of 20."

Hiding Caught Fruit

Finally, let's add the blocks that will make the fruit ImageSprites disappear after they collide with PickerSprite.

Click **FruitSprite1** and, when the blocks for the component appear, drag the **whenFruitSprite1.CollidedWith** block ❶ to the Viewer. Then, click **Fruit Sprite1**, drag the **setFruitSprite1.Visibleto** block ❷ to the Viewer, and snap it inside the **whenFruitSprite1.CollidedWith** block next to the word do. Next, in the Blocks pane, click the **Logic** blocks drawer, drag the **false** block ❸ to the Viewer, and snap it to the right side of the **setFruitSprite1.Visibleto** block.

Now duplicate these blocks for FruitSprite2, taking care to use the drop-down arrows both places you see FruitSprite1 to change to **FruitSprite2**. Then duplicate the blocks again, and be sure to change to **FruitSprite3**. These three event handlers hide the fruit ImageSprites when they hit PickerSprite so it looks like the picker successfully caught the fruit in the bucket. In the next chapter, you'll learn how to eliminate these duplicate blocks and accomplish the same task using a more sophisticated programming structure.

Now, following the plan for step 5, each time a fruit ImageSprite touches PickerSprite, the app should play a sound, increase the player's score by 1, display the score and the total pieces of fruit, and hide the fruit ImageSprite.

It's time to test the completed game! Open the app on your device, and you should see StartBtn at the top of the screen. Click it, and when it disappears, you should see the other game components appear. Now the fruit starts to randomly drop, and you can click LeftBtn and RightBtn to move PickerSprite back and forth across the screen to try to catch it.

Whenever PickerSprite catches a piece of fruit, you should hear a sound and see your score increase. If you placed your blocks correctly, the game should work as described, and you'll have successfully created the "Fruit Loot" game!

SUMMARY

In this chapter, you built the animated "Fruit Loot" app, a game where a player moves a fruit picker back and forth across the screen and earns points when the picker catches rapidly and randomly dropping fruit.

You learned how programmers animate an object by moving its x- and y-coordinates; use pseudorandom number generators to add randomness in games, simulators, and other applications; and work with arithmetic operators to manipulate data. You also practiced declaring and initializing variables to store and change information, and you learned how to provide required arguments for built-in methods with parameters.

In the next chapter, you'll do more with math operators and random number blocks and begin to make selections in your code using Control blocks. You'll use those tools to create part 1 of the "Multiplication Station" quiz app, which generates random, timed multiplication problems, evaluates solutions the user inputs, and then speaks to declare those answers right or wrong.

ON YOUR OWN

Save new versions of "Fruit Loot" as you modify and extend it working on these exercises. You can find solutions online at *https://nostarch.com/programwithappinventor/*.

1. Change the app so that it calculates and keeps track of how many pieces of fruit the picker fails to catch during a game. How can you calculate, store, and display this information using the existing event handlers and adding the smallest number of additional blocks?

2. Extend the game so that the frustrated owner of the fruit trees, who can't keep the fruit from falling over the fence, drops rocks down the fence to try to keep the picker from attempting to catch the falling fruit. Reduce the player's score each time the rock hits another sprite. What components and blocks will you add?

3. Extend the game even further so that the score label displays the number of times the rock hits another sprite.

MULTIPLICATION STATION: MAKING DECISIONS WITH CODE

In programming, you can use three structures to control the flow of a program: sequence, selection, and repetition. With the *sequence structure*, which is the default and simplest of the three options, code is executed in order, line by line. So far you've been using the sequence structure to create apps.

Unlike the sequence structure, the *selection* or *conditional structure* enables an app to make decisions by testing for a condition with a Boolean expression. A Boolean expression evaluates to a Boolean value, using *relational* or *comparison* operators like =, ≤, and ≥, to test the conditions. Because a Boolean value can only be either true or false, we execute one sequence of code if the condition is true and another if it's false.

The *repetition structure* also tests a Boolean condition and runs a sequence of code repeatedly while the condition is true, which is called *looping*. We'll use the repetition structure in Chapter 6.

THE SELECTION CONTROL STRUCTURE

In this chapter, we'll program apps to make decisions by comparing conditions using the if then blocks in the Control blocks drawer. We'll combine them with comparison operator blocks from the Logic and Math drawers to compare values. Then, we'll execute different blocks of code depending on whether the comparisons are true or false.

USING AN IF THEN BLOCK IN THE "FRUIT LOOT" GAME

In the "Fruit Loot" game in Chapter 3, we could have used an if then block to streamline some of our code. Remember that we programmed the action the app should take when an ImageSprite hits the picker, as shown here.

We should have included code in this event handler to make any sprite that collided with the picker disappear. But we couldn't, because we didn't yet know how to use the selection control structure to figure out exactly which sprite hit the picker and hide it.

Instead, we repeated CollidedWith event handlers for each fruit ImageSprite.

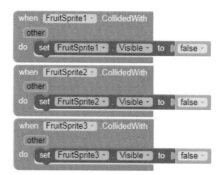

That's a lot of blocks! With the selection control structure, however, we don't have to use three duplicate fruit ImageSprite CollidedWith event handlers to figure out which fruit ImageSprite hit the picker and make it disappear.

Here is the *pseudocode*, or simple, plain English version of the code that determines which sprite hit the picker:

```
If (the sprite that hit PickerSprite is FruitSprite1)
    then (hide FruitSprite1).
If not, if (the sprite that hit PickerSprite is FruitSprite2)
    then (hide FruitSprite2).
If not,
    (hide FruitSprite3).
```

To re-create this pseudocode, we'll add an if then block to the picker ImageSprite's CollidedWith event handler, as shown in Figure 4-1.

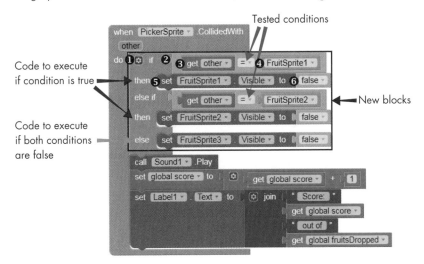

Code to execute if condition is true

Code to execute if both conditions are false

Tested conditions

New blocks

Figure 4-1: "Fruit Loot" blocks to find and hide the ImageSprite that hits the picker

To add the new blocks, go to the "Fruit Loot" game Blocks Editor and place the **if then** block ❶ inside the **whenPickerSprite.CollidedWith** block next to the word do.

You should see that the default if then block allows only one if and one then input. To add more inputs, click the blue mutator icon to the left of if, as shown in Figure 4-2.

In the dialog that opens, drag an **else if** block and then an **else** block into the **if** block.

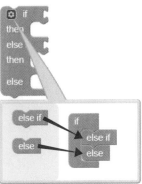

Figure 4-2: Adding else if and else sockets to the if then block

Checking Whether Colliding Fruit Is FruitSprite1

The PickerSprite's CollidedWith event handler provides the other event parameter, which represents the fruit ImageSprite that collided with the picker. Whenever an ImageSprite hits the picker, you compare other to FruitSprite1 to determine if FruitSprite1 is the sprite that hit the picker. If so, the blocks set the Visible property of FruitSprite1 to false to make it disappear.

To place the blocks for this behavior, from the Logic blocks drawer, drag in an = comparison operator block ❷ next to the word if. To fill the = block's operand sockets, mouse over the other event parameter above the mutator icon until you see a get other block ❸, then drag that block into the left socket. Then click FruitSprite1 in the Blocks pane and drag the FruitSprite1 block ❹ into the right socket. Next, in the Blocks pane, click FruitSprite1 again and drag in the setFruitSprite1.Visibleto block ❺ next to the word then. Finally, snap a false block ❻ from the Logic blocks drawer to the right of the setFruitSprite1.Visibleto block.

Checking Whether Colliding Fruit Is FruitSprite2

If FruitSprite1 is not the fruit that hit the picker, we check for our second condition by comparing other to FruitSprite2. Copy the blocks at ❷ through ❹ in Figure 4-1, place the duplicate blocks next to the words else if, and use the drop-down arrow to change FruitSprite1 to FruitSprite2 in the right operand socket of the duplicate = comparison operator block.

Then copy the blocks at ❺ and ❻, place the duplicate blocks next to the word then, and use the drop-down arrow to change FruitSprite1 to FruitSprite2 in the duplicate setter block. Now, when an ImageSprite other than FruitSprite1 hits the picker, these blocks compare other to FruitSprite2 to determine if FruitSprite2 is the sprite that hit the picker. If so, these blocks set its Visible property to false to make it disappear.

Checking Whether Colliding Fruit Is FruitSprite3

Finally, copy the blocks at ❺ and ❻ again, place the duplicate blocks next to the word else, and use the drop-down arrow to change FruitSprite1 to FruitSprite3 in the duplicate setter block. These last two blocks should execute only if neither FruitSprite1 nor FruitSprite2 matches other as the fruit ImageSprite that collided with the picker. In that case, FruitSprite3, the only sprite left, must have hit the picker and will disappear.

Instead of creating three different CollidedWith event handlers for the fruit ImageSprites, we refactor our code and use one if then block to handle multiple conditions. You now can delete the three original CollidedWith event handlers for the fruit ImageSprites, and, when you test the "Fruit Loot" app, it should work exactly the same.

BUILDING THE "MULTIPLICATION STATION" APP

Now that you've seen how if then blocks work, let's create the "Multiplication Station" quiz app. To build the app, you'll use if then blocks, along with the

Clock timer, random numbers, and arithmetic operators to generate multiplication problems. You'll also learn how to create an app that has more than one screen and experiment with the TextToSpeech and Notifier components to communicate with app users.

Log into App Inventor, create a new project, name it **MultiplicationStation**, and click **OK**.

DECOMPOSING "MULTIPLICATION STATION"

We can decompose the "Multiplication Station" app into five steps:

1. When the user opens the app, play a welcome message and show the start button.
2. When the user clicks the start button, open the practice screen.
3. When the Clock timer fires, display a random multiplication problem and an empty text box for the user to input the answer, and add 1 to the total number of problems.
4. When the user clicks the check answer button, check the answer and indicate whether it's right or wrong if an answer exists. If no answer exists, display an alert.
5. When the user clicks the end button, stop displaying problems and change the user interface to show only the final numeric and percentage practice scores.

You'll need the following components:

- **Button** (3) for the user to click to manually start and end practice and check answers
- **Clock** to fire after the user clicks the start button to display math problems at a 5-second interval
- **HorizontalArrangement** to hold the answer TextBox, the check answer button, and the answer result Label
- **Label** (3) to display math problems, answer results, and scores
- **Notifier** to display an alert to the user
- **Screen** (2) for the welcome screen and the practice screen
- **TextBox** for the user to input answers
- **TextToSpeech** (2) to say the welcome message and declare answers right or wrong
- Variable (7) to store problems, answers, and scores
- **VerticalArrangement** to hold all practice screen user interface components

Now let's lay out the app in the Designer.

LAYING OUT "MULTIPLICATION STATION" IN THE DESIGNER

Rather than briefly showing and hiding our introductory information like we did in Chapter 2's "Practice Makes Perfect" app, we'll use two different screens in this app: one to welcome the user and the other to display problems and evaluate the user's answers.

SETTING UP THE WELCOME SCREEN

To lay out Screen1, the welcome screen, drag a **Button** from the User Interface drawer and a non-visible **TextToSpeech** component from the Media drawer.
Figure 4-3 shows what the welcome screen should look like.

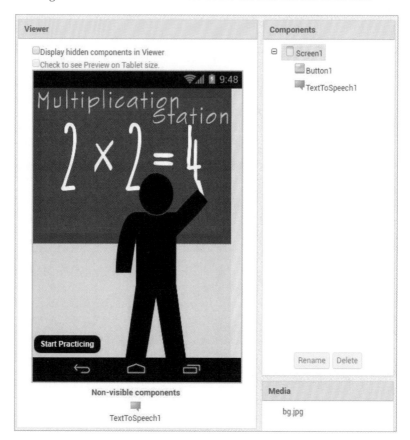

Figure 4-3: The Viewer, Component, and Media panes after laying out Screen1 of "Multiplication Station"

In the Components pane click **Screen1**. Then, in the Properties pane, adjust its vertical alignment so the Button sits at the bottom of the screen by clicking the drop-down arrow under **AlignVertical** and selecting **Bottom: 3**.

Then add a background image to the screen by clicking the text box under BackgroundImage and following the image upload instructions outlined in "Uploading a Picture" on page 27. Finally, remove the check marks under ShowStatusBar and TitleVisible to hide the status bar and the title of the screen, Screen1, when the app runs.

Now, let's style the Button that starts the app. Click Button1 in the Components pane. Then, in the Properties pane, change the background color to black by clicking **Default** under BackgroundColor and then clicking **Black** when the color list dialog opens. Next, bold its text by clicking the checkbox under FontBold, and change its shape by clicking the drop-down arrow under Shape and selecting **rounded**. Next, input Start Practicing in the text box under **Text**. To make the text white, click **Default** under TextColor and then **White** when the color list dialog opens.

Now let's create Screen2, where most of your app's action will happen.

CREATING THE PRACTICE SCREEN

To create Screen2, click the **Add Screen...** button above the Viewer. When the New Screen dialog opens, click **OK**, and you'll be taken to a brand new screen.

Now, in the Properties pane, we'll center Screen2 horizontally and vertically so that all components we place on it will be centered. To do this, click the drop-down arrows under both AlignHorizontal and AlignVertical and select **Center**. Next, add a background image to the screen by clicking the text box under BackgroundImage and following the image upload procedure. Finally, click the checkbox under TitleVisible to remove the check mark and keep the title of the screen from showing at the top of the app when it runs.

Problem Display

Now let's set up the components the user can see and interact with, by dragging the components from the Palette pane in the order that we want them to appear onscreen. Let's also adjust these components as we add them.

First, drag a **VerticalArrangement** from the Layout drawer, and center its contents horizontally and vertically by clicking the drop-down arrows under both AlignHorizontal and AlignVertical and selecting **Center**. Then, change the height of VerticalArrangement1 so it takes up half the vertical screen. To do so, click the text box under **Height**, and, when the dialog opens, input 50 in the text box next to the word percent and click **OK**. Finally, set its **Width** to **Fill parent**.

Now we'll drag all of the remaining visible components into Vertical Arrangement1. Drag a Label from the User Interface drawer into **Vertical Arrangement1**, click the **Rename** button to rename it as ProblemLbl, and click **OK**. Next, click the **FontBold** checkbox to make the text on ProblemLbl bold, input 25 in the **FontSize** text box to make the text larger, and replace the default Text for Label1 by inputting Problems appear here in the **Text** text box.

Answer Text Box and Check Answer Button

Next, drag a HorizontalArrangement from the Layout drawer into Vertical Arrangement1 under ProblemLbl. Select **Center** in both its AlignHorizontal and AlignVertical drop-down boxes and set its **Width** to **Fill parent**.

Now, drag three components from the User Interface drawer into HorizontalArrangement1. First, drag in a TextBox and rename it AnswerBox. Make its width 100 pixels by clicking the text box under **Width**, inputting 100 in the text box to the left of the word pixels, and clicking **OK**. Next, replace the Hint for TextBox1 in the **Hint** text box by entering Type your answer. Although you can't see this text in the Viewer, the user should see the hint in AnswerBox on the screen. Finally, click the checkbox under NumbersOnly, which will allow users to enter only numbers into AnswerBox.

Drag a Button into HorizontalArrangement1 to the right of AnswerBox and rename it CheckAnswerBtn. Then replace the default Text for Button1 on CheckAnswerBtn by inputting Check Answer in the **Text** text box. This is the button users will click to check their answers.

Finally, drag a Label into HorizontalArrangement1 to the right of Check AnswerBtn, rename it ResultLbl, and delete the default Text for Label1 from its **Text** text box. This label will remain invisible until it displays "Correct!" or "Incorrect" answer results.

End and Score Button

To display the score and percentage of answers correct, place a Label into VerticalArrangement1 under HorizontalArrangement1, and rename it ScoreLbl. Make the text bold and italic by clicking the checkboxes under FontBold and FontItalic and larger by entering 20 in the **FontSize** text box. Also, delete the default Text for Label1 from the **Text** text box so no text will show when the screen opens.

To create a button users can click to stop practicing and see their final scores, drag another Button into VerticalArrangement1 under ScoreLbl and rename it EndBtn. Style it exactly like Button1 on Screen1. Change the background color to black by selecting **Black** under BackgroundColor, make its text bold by clicking the checkbox under FontBold, and change its shape by clicking the drop-down arrow under **Shape** and selecting **rounded**. Then, enter End + Get Percentage in the text box under **Text**, and make the text white by selecting **White** under TextColor.

Finally, let's drag in the non-visible components: a Notifier component from the User Interface drawer, another TextToSpeech component from the Media drawer, and a Clock component from the Sensors drawer. Change the Clock's timer interval to 5 seconds by replacing the default 1000 with 5000 in the **TimerInterval** text box so a new problem will display every 5 seconds.

Now, in the Viewer pane, all visible components should show on Screen2, and the non-visible components—Notifier1, TextToSpeech1, and Clock1—should show under the screen. In the Components and Media panes, you should see a list of all components and images added, as shown in Figure 4-4.

Figure 4-4: The Viewer, Component, and Media panes after laying out Screen2 of "Multiplication Station"

Now that you've laid out all components, you're ready to program the app!

PROGRAMMING "MULTIPLICATION STATION"

For "Multiplication Station," we'll need five event handlers for the two screens. Three respond to events generated by the user's button clicks. One directs the app's action after a timer goes off at the time interval we've set. The other triggers action when the screen initializes or opens. We'll also create seven variables to store important information, and we'll use an if then block that has another if then block nested within it.

Click the **Blocks** button to switch to the Blocks Editor to begin programming.

STEP 1: PLAYING THE WELCOME MESSAGE

Let's start by telling the app what to do when Screen1 opens—namely, we want the TextToSpeech component to say the app's welcome message. The following code shows the event handler to program this behavior in step 1.

Switch back to Screen1 by selecting **Screen1** in the drop-down box to the right of the project's name. In the Blocks pane, click **Screen1** and drag the whenScreen1.Initialize event handler block ❶ to the Viewer. Then, click TextToSpeech1 in the Blocks pane and drag its callTextToSpeech1.Speak built-in method block ❷ next to the word do. Next, in the Blocks pane, click the **Text** blocks drawer and drag an empty string block ❸ (the first block in the drawer), next to the word message, into the socket for the method's message parameter.

The message parameter is required by the TextToSpeech1 Speak method in order for the app to speak. Add that message by clicking the empty string block's text area and entering the following: Welcome to Multiplication Station! You will have 5 seconds to answer each problem. Click the Start Practicing button to begin. Altogether, these blocks program step 1 of the app, greeting the user with a spoken welcome message when the app opens.

STEP 2: MOVING TO THE PRACTICE SCREEN

In addition to hearing the welcome message, users should see Button1, which they can click to display the multiplication problems in Screen2. The following blocks handle this step.

Click **Button1** in the Blocks pane and drag the whenButton1.Click block ❶ to the Viewer. Place the **open another screen** block ❷ inside the whenButton1 .Click block next to the word do. From the Text blocks drawer, drag in another empty string block ❸ and snap it onto the **open another screen** block. Enter **Screen2** (the exact name of the screen we want to open) inside the empty string block. Now when the user clicks Button1, the app should open Screen2 just like we planned.

To see how these blocks work, switch back to **Screen1** and let's live-test with a device, as outlined in "Live-Testing Your Apps" on page xxii. Once you click **Connect ▸ AI Companion** in the top menu bar and scan the QR code with your phone's AI2 Companion app, the "Multiplication Station" app should open on your phone. As long as the blocks are placed correctly, you should hear the welcome message and, after clicking Button1, you should see Screen2. For now, nothing else should happen. Close AI2 Companion for now, and we'll reconnect when we test again.

STEP 3: DISPLAYING RANDOM MULTIPLICATION PROBLEMS

Let's now program step 3 of the app, where we display a random multiplication problem every 5 seconds. When this happens, we'll count each problem to keep a running total. Make sure you're in the Blocks Editor for Screen2 to program this step.

Setting Global Variables

We'll use three global variables in this step: a and b to store the numbers in the multiplication problems and problems to store the total number of problems. We'll create and initialize the global variables as shown here.

① initialize global ⓐ to ②ⓞ ① initialize global ⓑ to ②ⓞ ① initialize global ⓟproblems to ②ⓞ

For each variable, click the **Variables** block drawer and drag an initialize global name block ❶ to the Viewer, click name, and replace it with the name of the variable. Then drag a 0 number block ❷ from the Math drawer and snap it onto the initialize global block. Because all three variables are global, they can be used by all our event handlers and they stand alone in our code, outside of all our event handler blocks.

Choosing Random Numbers to Multiply

Next, we'll program the Clock's timer. Because we kept the default setting for the Clock's TimerEnabled property in the Designer, the timer should automatically begin firing at 5-second intervals once Screen2 opens. The following event handler blocks tell the app what to do each time the timer fires.

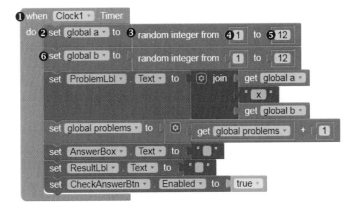

In the Blocks pane, click **Clock1** and, when the blocks for the component appear, drag the **whenClock1.Timer** block ❶ to the Viewer. Next, in the Blocks pane, click the **Variables** blocks drawer, drag the **set global a to** block ❷ to the Viewer, and snap it inside the `whenClock1.Timer` block next to the word do. Then click the **Math** blocks drawer and drag a `random integer` block ❸ to the right of the `set global a to` block.

The random integer block will generate a random integer between a specified range of numbers. To set this range, leave the 1 in the left socket of the random integer block as is ❹, click the 100 in the right socket, and replace it by entering 12 ❺.

So far we've set the value for variable a, the random number on the left side of each multiplication problem that will show in the app. To set the number for variable b, the right side of the multiplication problems, copy the blocks at ❷ through ❺, place the duplicate blocks under the original, and use the drop-down arrow to change a to b ❻ in the duplicate `set global` block.

Displaying the Problems

Now we'll add the blocks to the `Timer` event handler that display the multiplication problems to the user in `ProbemLbl`.

In the Blocks pane, click `ProbemLbl`, drag the `setProblemLbl.Textto` block ❶ to the Viewer, and snap it inside the `whenClock1.Timer` block under the set global b to block. Then click the **Text** blocks drawer, drag a `join` block ❷ to the Viewer, and snap it to the right of the `setProblemLbl.Textto` block. This allows us to join the random number values of the a and b variables and the multiplication operator, ×, into one string. For example, if a = 3 and b = 6, we want to join them with the operator to display 3 × 6.

First, we'll add another string input to the two default inputs in the `join` block. To do this, click the blue mutator icon to the left of the word `join`, and drag one additional string block to the **join** block in the dialog that opens.

Then, for the `join` block's top input, click the **Variables** blocks drawer and drag in the `get global a` block ❸. For the `join` block's second input, drag in an empty string block ❹ from the Text blocks drawer, click the empty string block's text area, and enter **x** (with spaces before and after the x). Then, for the `join` block's bottom input, click the **Variables** blocks drawer again and drag in the `get global b` block ❺. Now, every 5 seconds, these blocks will get new random values for a and b, join them with the multiplication operator, and display the multiplication problem in `ProbemLbl`.

Counting the Number of Problems Displayed

To keep a running total of the number of problems, we need the app to count each problem as it's displayed. The next blocks in the `Timer` event handler program this.

Click the **Variables** blocks drawer, drag the `set global problems to` block ❶ to the Viewer, and snap it inside the `whenClock1.Timer` block under the `setProblemLbl.Textto` block. Then drag an addition operator block ❷ from the Math drawer and snap it to the right of the `set global problems to` block. Now click the **Variables** blocks drawer again and drag the `get global problems` block ❸ into the addition block's left operand socket and a **1** number block ❹ from the Math drawer into its right operand socket. These blocks keep a running total of the problems displayed by adding 1 to the current value of the `problems` variable every 5 seconds, each time a new multiplication problem appears onscreen.

Refreshing the AnswerBox, ResultLbl, and CheckAnswerBtn

We need the app to do three more things each time the `Timer` fires and it displays a new math problem:

1. Erase any text left in `AnswerBox` so the user has a clean box to input answers
2. Remove the "Correct!" or "Incorrect" text displayed in response to any prior answer
3. Enable the `CheckAnswerBtn`, which we'll later program the app to temporarily disable after the user clicks it

The following blocks accomplish these three tasks.

Let's prepare the app to accept an answer to a new problem and equip its `CheckAnswerBtn` to check that answer. Click **AnswerBox** in the Blocks pane, drag its `setAnswerBox.Textto` block ❶ to the Viewer, and snap it inside the `whenClock1.Timer` block under the set `global problems to` block. Then click the **Text** blocks drawer, drag in an empty string block ❷, and snap it to the right of the `setAnswerBox.Textto` block.

Click **ResultLbl** in the Blocks pane, drag its `setResultLbl.Textto` block ❸ to the Viewer, and snap it inside the `whenClock1.Timer` block under the setAnswer Box.Textto block. Then click the **Text** blocks drawer, drag in another empty string block ❹, and snap it to the right of the `setResultLbl.Textto` block.

Click **CheckAnswerBtn** in the Blocks pane, drag its `setCheckAnswerBtn.Enabledto` block ❺ to the Viewer, and snap it inside the `whenClock1.Timer` block under the setResultLbl.Textto block. Then click the **Logic** blocks drawer, drag in a **true** block ❻, and snap it to the right of the `setCheckAnswerBtn.Enabledto` block.

Now live-test to see how these blocks work. If the AI2 Companion button is grayed out when you attempt to reconnect the AI2 Companion app, click **Connect ▸ Reset Connection** in the top menu bar. Then, click **Connect ▸ AI**

Companion and scan the QR code with your device's AI2 Companion app. You should still see Screen2 since you're connecting AI2 Companion while working on your Screen2 blocks.

Five seconds after the app opens, you should see a multiplication problem appear in place of "Problems appear here" and then see a new problem every 5 seconds after that. Also, if you enter numbers in AnswerBox, they should disappear after 5 seconds. If you don't see any multiplication problems, the problems don't show as desired, or the answers don't disappear after each 5-second interval, debug your code and test again.

Next, we'll tell the app what to do when the user clicks the Check Answer button.

STEP 4: CHECKING ANSWERS

Let's program step 4 of the app, where the user clicks CheckAnswerBtn. First, we want to determine whether the user has entered anything into AnswerBox. If so, we'll program the app to evaluate the answer and indicate whether it's wrong or right by speaking and displaying the result in a label. If the answer's right, the app increases the score by 1. If, on the other hand, the user clicks CheckAnswerBtn without entering anything into AnswerBox, the app displays an alert. In this step, you're *validating* user input, which means checking whether the user has entered the type of data required. Programmers commonly validate user form input and use it only if it's valid or notify the user if it's not.

We'll start our code for this step by creating and initializing three more global variables. Create the variables answer, c, and score. Then initialize answer to the value of an empty string block from the Text blocks drawer, and c and score to the value of 0. The variables should look as shown here.

We'll use these variables to store the user's answer (answer), the correct answer (c), and the user's score (score).

Testing Our First Condition: Checking Whether an Answer Exists

Let's now program the CheckAnswerBtn event handler, which includes two if then blocks, one nested within the then socket of the other. This means that if the first if then condition is true, the app will need to test another condition. The first if then block tests whether the user has entered any numbers into AnswerBox, providing one set of instructions for the app to follow if the user has entered numbers and different instructions if the box is empty.

If the user has entered an answer, the second if then block tests the answer and then tells the app what to do depending on whether it is right or wrong. Here is the code for programming this nested conditional.

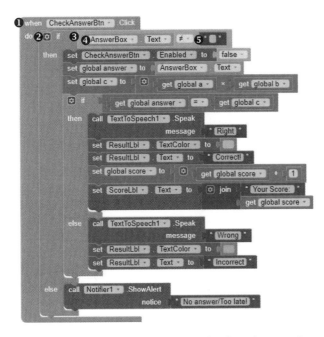

To place the blocks that test whether the user has entered an answer, click **CheckAnswerBtn** in the Blocks pane and drag the **whenCheckAnswerBtn.Click** block ❶ to the Viewer. Then, in the Blocks pane, click the **Control** blocks drawer, drag an **if then** block ❷ to the Viewer, and snap it inside the **when CheckAnswerBtn.Click** block next to the word **do**. Since we'll need an **else** statement for this conditional, click the **if then** block's blue mutator icon, and drag an **else** block into the **if then** block in the dialog that opens.

Then, from the Logic blocks drawer, drag in an **=** comparison operator block ❸ next to the word **if**, and click the drop-down arrow by the **=** sign to select the not equal sign, **≠**. To fill the **≠** block's operand sockets, click **Answer Box** in the Blocks pane and drag its **AnswerBox.Text** block ❹ into the left socket, then click the **Text** blocks drawer and drag an empty string block ❺ into the right socket. We've now set up the test condition that checks whether **Answer Box** is empty.

Preparing to Evaluate the User's Answers

Now we'll tell the app what to do if **AnswerBox** isn't blank (we'll program what the app should do if it is blank later in this chapter). If **AnswerBox** isn't blank, we want the app to disable **CheckAnswerBtn** so the user can't click it again while the app's checking the current answer. Then we want the app to set the value of the global **answer** variable to the number the user has entered into **AnswerBox** so the app can check it later, and to set the value of the global **c** variable to the value of global **a** multiplied by global **b**, which is the current problem's correct answer. The following blocks program these three actions.

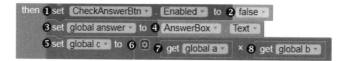

Click **CheckAnswerBtn** in the Blocks pane, drag its `setCheckAnswerBtn.Enabledto` block ❶ to the Viewer, and snap it into the `if then` block next to the word then. Then click the **Logic** blocks drawer, drag in a `false` block ❷, and snap it to the right of the `setCheckAnswerBtn.Enabledto` block. This prevents the user from clicking `CheckAnswerBtn` again while the app's checking an answer.

Next, click the **Variables** blocks drawer and drag the `set global answer to` block ❸ into the `if then` block under the `setCheckAnswerBtn.Enabledto` block. Then click **AnswerBox** in the Blocks pane, drag in its **AnswerBox.Text** block ❹, and snap it to the right of the `set global answer to` block. This sets the value of the answer variable to the number the user entered into `AnswerBox` so we can compare that number with the correct answer.

Now, click the **Variables** blocks drawer and drag the `set global c to` block ❺ into the `if then` block under the `set global answer to` block. Then click the **Math** blocks drawer, drag in the multiplication operator block ❻, and snap it to the right of the `set global c to` block. To fill the multiplication block's operand sockets, click the **Variables** blocks drawer and drag the `get global a` block ❼ into the left socket and the `get global b` block ❽ into the right socket. These blocks set the value of variable c to the value of variable a multiplied by the value of variable b, which is the correct answer to the problem.

So far, the blocks say: if `AnswerBox` isn't blank, disable `CheckAnswerBtn`, then set the value of answer to the number entered by the user, and set the value of c to the correct answer.

Testing Our Second Condition: Evaluating the Answers

Now we'll tell the app to compare answer and c and then do one thing if they are equal, which means the user's answer to the problem is correct, and another thing if they aren't equal, which means the user's answer is wrong.

We'll start by placing the blocks that set up the second test condition.

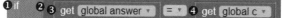

In the Blocks pane, click the **Control** blocks drawer and drag another `if then` block ❶ to the Viewer inside the first `if then` block under the set global c to block. Then, click this new `if then` block's blue mutator icon to the left of the word if, and drag an **else** block into the `if then` block in the dialog that opens.

Then, from the Math blocks drawer, drag in an = comparison operator block ❷ next to the word if. To fill the = block's operand sockets, click the **Variables** blocks drawer and drag the `get global answer` block ❸ into the left socket and the `get global c` block ❹ into the right socket. These blocks set up our second test condition and check whether the user's answer matches the correct answer.

Handling Correct Answers

The next blocks tell the app what to do if the value of the two variables is the same, meaning the user's answer is correct.

Click **TextToSpeech1** in the Blocks pane and drag its `callTextToSpeech1` `.Speak` block ❶ into the second `if` `then` block next to the word then. Set the argument for its message parameter to `Right` by snapping an empty string block ❷ from the Text drawer next to the word message, clicking the string block's text area, and entering `Right`.

Next, click **ResultLbl** in the Blocks plane and drag its `setResultLbl.Text` `Colorto` block ❸ into the second `if` `then` block under the `callTextToSpeech1` `.Speak` block. Then click the **Colors** blocks drawer, drag in the green color block ❹, and snap it to the right of the `setResultLbl.TextColorto` block.

Now click **ResultLbl** in the Blocks plane again and drag its `setResultLbl` `.Textto` block ❺ into the second `if` `then` block under the `setResultLbl.Text` `Colorto` block. Click the **Text** blocks drawer, drag in an empty text string block ❻, snap it to the right of the `setResultLbl.Textto` block, and enter `Correct!` into the text string block's text area. If the user's answer is correct, these blocks direct the app to say so and show the result in green in the `Label`.

Let's pause now to live-test again. If you've placed the blocks as shown, after you enter correct answers to problems and click `CheckAnswerBtn`, you should hear the device say "Right" and see the word "Correct!" in green text to the right of `CheckAnswerBtn`. Also, if you try to click `CheckAnswerBtn` again before the app checks your answer, the button shouldn't work. Nothing should happen yet when you enter an incorrect answer. Make sure this part of the app is working correctly before continuing.

Increasing the Score

The next blocks increment and display the user's score, which means we first need the app to count each answer the user gets right.

To program this, click the **Variables** blocks drawer, drag the `set global` `score to` block ❶ to the Viewer, and snap it inside the second `if` `then` block under the `setResultLbl.Textto` block. Then drag an addition operator block ❷ from the Math drawer and snap it to the right of the `set global score to` block. Now click the **Variables** blocks drawer again, and drag the `get global` `score` block ❸ into the addition block's left operand socket and a `1` number block from the **Math** drawer ❹ into its right operand socket. These blocks add 1 to the current value of the score variable each time the user answers a problem correctly.

The next blocks display the user's score in ScoreLbl. In the Blocks pane, click ScoreLbl, drag the setScoreLbl.Textto block ❺ to the Viewer, and snap it inside the second **if then** block under the set global score to block. Then click the **Text** blocks drawer, drag a **join** block ❻ to the Viewer, and snap it to the right of the setScoreLbl.Textto block.

For the **join** block's top input, drag in an empty string block ❼ from the Text blocks drawer, click the string block's text area, and enter **Your Score:** (including a space after the colon). Then, for the **join** block's bottom input, click the **Variables** blocks drawer and drag in the **get global score** block ❽.

Let's pause here to live-test again. If you've placed the blocks as shown, after you enter correct answers to problems and click CheckAnswerBtn, you should see the score display under AnswerBox in ScoreLbl. Nothing should happen yet when you enter an incorrect answer. Make sure this part of the app is working correctly before you move on.

Handling Incorrect Answers

The blocks in the prior section told the app what to do when the user answers a problem correctly. The blocks shown here complete the else portion of the second if then else block and guide the app's behavior when the user's answer is wrong.

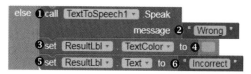

First we'll place the blocks that make the app speak to tell the user the answer is wrong. Click TextToSpeech1 in the Blocks pane and drag its callTextToSpeech1.Speak block ❶ into the second **if then** block next to the word else. Set its message to "Wrong" by snapping an empty string block ❷ from the Text drawer next to the word message, clicking the string block's text area, and entering **Wrong**.

Next, click ResultLbl in the Blocks plane and drag its setResultLbl.Text Colorto block ❸ into the second **if then** block under the second callTextTo Speech1.Speak block. Then click the **Colors** blocks drawer, drag in the pink color block ❹, and snap it to the right of the setResultLbl.TextColorto block.

Then click ResultLbl in the Blocks pane again and drag its setResultLbl .Textto block ❺ into the second **if then** block under the second setResult Lbl.TextColorto block. Now click the **Text** blocks drawer and drag in an empty text string block ❻, snap it to the right of the setResultLbl.Textto block, and enter **Incorrect** into the text string block's text area. If the user's answer is incorrect, these blocks have the app say so and show that result in pink in ResultLbl.

Try live-testing now by entering an incorrect answer and clicking CheckAnswerBtn. If you've placed the blocks as shown, you should now hear the device say "Wrong" and see the word "Incorrect" in pink text to the right of CheckAnswerBtn. Make sure this part of the app is working correctly before continuing.

Handling an Empty Answer Box

Until now, the blocks for this step have told the app what to do when the user clicks the CheckAnswerBtn after entering a number into AnswerBox. Now, with the following blocks, we tell the app what to do when the user clicks CheckAnswerBtn when AnswerBox is empty.

Click **Notifier1** and drag its **callNotifier1.ShowAlert** built-in method block ❶ into the second else socket. Then, in the Blocks pane, click the **Text** blocks drawer and drag an empty string block ❷ into the socket for the method's notice parameter.

This notice parameter holds information the method needs in order to show an alert. Add that notice by clicking the string block's text area and entering **No answer/Too late!**. These blocks direct the app to display a small pop-up notice when the user clicks CheckAnswerBtn without entering an answer.

Now that you've fully programmed CheckAnswerBtn, let's live-test again. If you've placed the blocks as shown, once you've entered an answer and clicked CheckAnswerBtn, you should hear and see whether it's right or wrong, and you should see the score displayed under AnswerBox in ScoreLbl. If you click CheckAnswerBtn and haven't entered an answer in time, you should see a pop-up alert. If any part of this action is working incorrectly, debug your program. Once the Check Answer button is working correctly, close the AI2 Companion on your device, and let's move on to program the last part of the app: the event handler for EndBtn.

STEP 5: ENDING PRACTICE AND SHOWING THE FINAL SCORE

Let's now program step 5 of the app, where the user presses EndBtn to stop practicing and see the final score as well as the percentage of problems answered correctly. We'll create one additional global variable for this part, percentage, which you can initialize to the value of 0.

initialize global `percentage` to `0`

Now we can program the EndBtn event handler as shown here.

![Blockly blocks showing a when EndBtn.Click event handler]

❶ when **EndBtn** .Click
do **❷** set **Clock1** . **TimerEnabled** to **❸** false
    set **ProblemLbl** . **Visible** to false
    set **HorizontalArrangement1** . **Visible** to false
    set **EndBtn** . **Visible** to false
    set global **percentage** to ⚙ (get global **score** / get global **problems**)
                      × 100
    set **ScoreLbl** . **Text** to ⚙ join **ScoreLbl** . **Text**
                      " out of "
                      get global **problems**
                      " ("
                      get global **percentage**
                      "%)"

Once the user clicks the button, the first thing we need to do is stop new problems from showing every 5 seconds. To do this, click **EndBtn** in the Blocks pane and drag the **whenEndBtn.Click** block ❶ to the Viewer. Then click **Clock1** in the Blocks pane, drag the **setClock1.TimerEnabledto** block ❷ to the Viewer, and snap it inside the **whenEndBtn.Click** block next to the word do. Then, in the Blocks pane, click the **Logic** blocks drawer, drag the **false** block ❸ to the Viewer, and snap it to the right side of the **setClock1.TimerEnabledto** block. These blocks disable the Clock's timer so the app stops displaying new problems.

Hiding Elements We Don't Need

We also want ProblemLbl, HorizontalArrangement1 (which contains AnswerBox, CheckAnswerBtn, and ResultLbl), and EndBtn to disappear, since we no longer need them on the screen. Here are the blocks that hide components the user no longer needs.

❶ set **ProblemLbl** . **Visible** to **❷** false
❸ set **HorizontalArrangement1** . **Visible** to **❹** false
❺ set **EndBtn** . **Visible** to **❻** false

To program this behavior, click **ProblemLbl** in the Blocks pane, drag the **setProblemLbl.Visibleto** block ❶ to the Viewer, and snap it inside the **whenEnd Btn.Click** block under the setClock1.TimerEnabledto block. Then, in the Blocks pane, click the **Logic** blocks drawer, drag the **false** block ❷ to the Viewer, and snap it to the right side of the **setProblemLbl.Visibleto** block.

These blocks reset the Visible property for ProblemLbl to false so that no problem shows after the user clicks EndBtn. Now place the blocks at ❸, ❹, ❺, and ❻ to do the same for HorizontalArrangement1 and EndBtn, so that after clicking EndBtn, all the user will see on the screen is the score and percentage correct.

Calculating the Percentage of Correct Answers

The app needs to compute the percentage of correct answers before displaying it. The following blocks tell the app how to compute and store the user's percentage.

Click the **Variables** blocks drawer, drag the `set global percentage to` block ❶ to the Viewer, and snap it inside the `whenEndBtn.Click` block under the `setEndBtn.Visibleto` block. Then drag a multiplication operator block ❷ from the Math drawer and snap it to the right of the `set global percentage to` block.

Now click the **Math** drawer again, drag a division operator block ❸ into the multiplication block's first operand socket, and drag a `100` number block ❹ into the multiplication block's second operand socket. Return to the Variables blocks drawer and drag the `get global score` block ❺ into the division block's left operand socket and the `get global problems` block ❻ into its right operand socket. These blocks compute the percentage of problems answered correctly by dividing the user's score by the total number of problems and multiplying that result by 100.

Showing the Percentage of Correct Answers

Now we'll display that percentage for the user with the blocks shown here.

In the Blocks pane, click **ScoreLbl**, drag the `setScoreLbl.Textto` block ❶ to the Viewer, and snap it inside the `whenEndBtn.Click` block under the set `global percentage to` block. Then click the **Text** blocks drawer, drag a `join` block ❷ to the Viewer, and snap it to the right of the `setScoreLbl.Textto` block. Now click the `join` block's mutator icon to add another four string input sockets.

For the `join` block's first input, click **ScoreLbl** and drag in its `ScoreLbl` `.Text` block ❸. For the `join` block's second input, drag in an empty string block ❹ from the Text blocks drawer, click the string block's text area, and enter **out of** (including spaces before the word **out** and after the word **of**). For the `join` block's third input, click the **Variables** blocks drawer and drag in the `get global problems` block ❺.

For its fourth input, drag in another empty string block ➏ from the Text blocks drawer, click the string block's text area, and enter (, including a space before the parenthesis. For the fifth input, click the **Variables** blocks drawer again and drag in the `get global percentage` block ➐. Finally, for the `join` block's bottom input, drag in an empty string block ➑ from the Text blocks drawer, click the string block's text area, and enter %).

These blocks display the user's score and percentage correct onscreen. For instance, if the user's score is 20 and the total number of problems is 100, `ScoreLbl` should display "Your Score: 20 out of 100 (20%)".

TESTING THE APP

Now we can test the completed app! First, move back to `Screen1` by selecting it in the drop-down above the Blocks Viewer. Then, reset the connection to AI2 Companion by clicking **Connect ▸ Reset Connection** in the top menu bar and then **Connect ▸ AI Companion**. Next, scan the QR code with your device's AI2 Companion app.

When the app opens on your device, you should hear the welcome message. Click **Start Practicing** to move to the next screen. There you should see a new problem appear every 5 seconds and, when you click Check Answer, if your answer `TextBox` is not empty, you should hear and see whether your answer's right or wrong, and see the score displayed under `AnswerBox`. If you click Check Answer and haven't typed an answer, you should see a pop-up alert.

Now click `EndBtn` and all components should disappear from the screen except for `ScoreLbl`, which should correctly display your score and percentage of problems correct. If you placed the blocks correctly, the app should work as expected. You've successfully created the "Multiplication Station" quiz app!

SUMMARY

In this chapter, you built the "Multiplication Station" quiz app, where users answer random, timed multiplication problems, learn whether their answers are correct, earn points for correct answers, and view the score and percentage of correctly answered problems.

In the process of building this app, you learned how to use the sophisticated selection control structure, which controls the flow of a program by evaluating conditions and making decisions. To do this, you used App Inventor's Control blocks along with its pseudorandom number generator and Math and Logic operator blocks. You also created and manipulated data stored in several variables and learned to use the `TextToSpeech` and `Notifier` components to communicate with app users.

In the next chapter, you'll learn how to create and process lists using Lists blocks as we build the "Beat the Bus" app, which uses your device's GPS and location-sensing capabilities to display your school bus's location and text your friends or family with periodic updates along the route.

ON YOUR OWN

Save new versions of "Multiplication Station" as you modify and extend it for the following exercises. You can find solutions online at *https://nostarch.com/programwithappinventor/*.

1. Change the app so that if a user's answer is incorrect, new problems stop appearing and the app displays the correct answer, after which the user may continue the practice session. Where will the app display the correct answer? How will the user resume practice, if desired? Will you need another if then block, additional components, or more variables?
2. Extend the app so that the user can choose to practice multiplication or division. How will your algorithm change? What components and blocks will you need to add and modify?

5

BEAT THE BUS: TRACKING LOCATION WITH MAPS AND SENSORS

In Chapters 3 and 4, we used variables to store and process information, which allowed us to track and report data to users. In the "Fruit Loot" and "Multiplication Station" apps, we created variables that hold one value at a time, like the game score, number of fruits dropped, and percentage of multiplication problems correct.

But imagine how much more powerful our apps would be if we could store and manipulate multiple values in a single variable! For example, we could display a game's top 10 scores or allow users to choose from a list of 50 states, without needing to create a different variable for each score or state.

In most programming environments you can store multiple values in a data structure called an *array*. In App Inventor, you can store multiple values in variables using the built-in Variables, Lists, and Text blocks. The following variable stores multiple values as a list.

You can use Lists blocks to manipulate list variables by adding items to a list, searching inside a list, or removing items from a list. You can also access each list item by referring to its *index*, the number of its position in the list. For instance, the first item in the size list, S, is at index 1, the second item is at index 2, and so on.

App Inventor numbers indexes differently from most traditional programing languages, which number the first item in a list as index 0, the second item as index 1, and so on.

In this chapter, we'll work with four lists that contain *static* (unchanging) preset values. Three of the lists will be interrelated, so that you'll use an index of one list to access values in the two other lists. We'll also create our first local variable to simplify a somewhat complex event handler.

BUILDING THE "BEAT THE BUS" APP

In this chapter, along with the four lists just mentioned, you'll use ListPicker, LocationSensor, Texting, and some of App Inventor's newer Maps components to create the "Beat the Bus" app. This app allows your parents or friends to track your location without using the GPS and location-sensing functions on their phones, which may drain their batteries. Instead, this app relies on the location sensor on your own device.

Log into App Inventor, create a project named **BeatTheBus**, and click **OK**.

DECOMPOSING "BEAT THE BUS"

In "Beat the Bus," the user activates location tracking by selecting a telephone number and destination from preset lists. The app then sends periodic text messages to the selected telephone number, showing the user's location along the route.

We can decompose the app into the following steps:

1. When a user opens the app, display a Map showing the user's current location with a Marker pointing to the user's home. Show a ListPicker for the user to select a phone number. Once the user selects a number, show a ListPicker for the user to select a destination.

2. Before the user selects a destination, set the ListPicker options from a list of locations.

3. Once the user selects a destination, move the `Map` and `Marker` to the selected destination, and text the starting location and destination to the selected number.

4. When the user's location is 5 miles or less away from the selected destination, periodically text the user's location to the selected number.

You'll need the following components:

- Global variable (4) to store telephone numbers, destinations, latitudes, and longitudes
- `Image` for app design
- `ListPicker` (2) for the user to select from telephone number and destination lists
- Local variable to store distance
- `LocationSensor` to provide periodic data related to the user's location
- `Map` to show the user's geographic location and hold the `Marker`
- `Marker` to point to the user's selected destination
- `Texting` to send text messages

Now let's lay out the app.

LAYING OUT "BEAT THE BUS" IN THE DESIGNER

To make sure that all components in `Screen1` will be centered at the top of the screen, click the drop-down arrow under **AlignHorizontal** and select **Center: 3**. Next, change the screen's background color to yellow by clicking **Default** under **BackgroundColor** and then **Yellow** when the color list dialog opens. Finally, remove the check mark under **TitleVisible** to hide the title of the screen when the app runs.

Next, let's drag the necessary components from their drawers in the Palette pane onto the Viewer pane and adjust their properties as we go. First, drag in an **Image** from the User Interface drawer and set its **Height** to **Fill parent**. Then upload the picture to display by clicking the text box under **Picture** and following the image upload instructions outlined in "Uploading a Picture" on page 27. Finally, click the checkbox under **Scale PictureToFit** so the image will take up the entire width of the screen.

To display a `Map` centered at your user's starting location, place a `Map` component from the Maps drawer under `Image1`. Note that you must place a `Map` on the screen before you can add any other Maps component. Adjust the `Map`'s properties by replacing the coordinates in the **CenterFromString** text box to the *latitude* and *longitude* for your app's starting address, separated by a comma and space.

NOTE *You can locate any place on a map by its latitude and longitude coordinates. The* latitude *of a place is its distance north or south of the equator, while the* longitude *is its distance east or west of the prime meridian. You can find the latitude and longitude coordinates for a street address by entering it into the Latitude/Longitude Finder tool at* https://www.latlong.net/.

Then, to change the Map's height to 50 percent so it takes up half the vertical screen, click the text box under **Height**; when the dialog opens, input **50** in the text box next to the word percent and click **OK**. Next, set its **Width** to **Fill parent**, click the checkbox under **ShowUser** so an icon representing the user will show on the Map and track the user's location, and change the Map zoom level from the default 13 to 10 by inputting **10** in the ZoomLevel text box.

Now drag a **Marker** from the Maps drawer onto the Map. Adjust the Marker in the Properties pane by clicking the text box under **Latitude** and entering the latitude for the user's home address, then clicking the text box under **Longitude** and entering its longitude.

Next, to let users select a destination and a phone number to send it to, under **Map1**, drag two **ListPicker**s from the User Interface drawer to the Viewer. Rename **ListPicker1** to **NumberPicker** and **ListPicker2** to **PlacePicker**. In the Properties pane for both, change the background color to yellow by clicking **Default** under **BackgroundColor** and then **Yellow** when the color list dialog opens. Then make the text bold by clicking the checkbox under FontBold, and make the text larger by inputting **25** in the **FontSize** text box. Replace the default Text for ListPicker1 by entering **Select a Number** in the **Text** text box for NumberPicker and the default Text for ListPicker2 by entering **Select a Place** in the **Text** text box for PlacePicker.

Change the **Width** for both ListPickers to **Fill parent**, and change the color of the items in the selection drop-downs to yellow by clicking **Default** under **ItemTextColor** and then **Yellow** when the color list dialog opens. Then, in the Properties pane for **PlacePicker**, click the checkbox under **Visible** to remove the check mark and keep PlacePicker from showing when the app first opens. We'll program it later to appear after the user picks a phone number.

Finally, drag in the non-visible components: the LocationSensor component from the Sensors drawer and the Texting component from the Social drawer. In the Properties pane, change the **LocationSensor**'s **TimeInterval** from the default 60,000 milliseconds to 1,000 by choosing **1000** in the drop-down box below **TimeInterval**. This means that the app's LocationSensor will report location changes each second if location data is available and the location has changed since the last interval. We're making this change so that later you can program your app to text updated location information as frequently as every second after new data becomes available.

Now that we've added LocationSensor1, click the **Map** in the Components pane and, in its Properties pane, choose its associated LocationSensor by clicking **None** under **LocationSensor**, selecting **LocationSensor1** in the dialog that opens, and clicking **OK**.

Screen1 should look like Figure 5-1.

Figure 5-1: The Viewer, Component, and Media panes after laying out "Beat the Bus"

In the Viewer pane, you should see all visible components on the screen and the non-visible components under Screen1. In the Components pane, you should see a list of every component you've dragged onto the Viewer pane. In the Media pane, you should see the image you've uploaded.

PROGRAMMING "BEAT THE BUS"

Now that you've laid out all components, you're ready to program the app in the Blocks Editor. For "Beat the Bus," we'll program five event handlers. Four of these respond to events related to the two ListPickers. The other one directs the app's action each time the LocationSensor detects a location change. We'll also create four global list variables and one local variable to store important information, and we'll use two if then blocks to test conditions to direct the flow of the app's action.

Click the **Blocks** button to switch to the Blocks Editor, and let's start programming the app's four steps.

STEP 1: CHOOSING THE PHONE NUMBER TO TEXT

When the screen opens, we want the user to see the Map and Marker, both with the properties we set in the Designer. We also want to display NumberPicker, from which the user will select a telephone number for the app to text. Let's start our code by telling the app what to do before and after the user clicks NumberPicker and selects a number.

Creating a Global List Variable

We'll use a global list variable to store the telephone numbers in NumberPicker by creating and initializing the phoneNumbers variable.

Click the **Variables** block drawer and drag an `initialize global name` block ❶ to the Viewer, click `name`, and replace it with **phoneNumbers**. Then drag a `make a list` block ❷ from the Lists drawer and snap it to the right side of the `initialize global phoneNumbers` block.

Now, in order to add three telephone numbers to the list, we need to add another list item input to the `make a list` block's two default inputs. Click the blue mutator icon to the left of the words `make a list`, and drag one additional `item` block to the `make a list` block in the dialog that opens.

Then, drag three empty string blocks ❸ from the Text drawer and snap them into the sockets of the `make a list` block. Enter a telephone number into each empty string block, without any dashes or spaces (if you don't have a phone number for any of the empty strings, enter **1111111**). Together, these blocks create the global phoneNumbers variable, which we can use in all our event handlers.

Handling NumberPicker's BeforePicking Event

Now that we've created phoneNumbers, we have the data we need for NumberPicker. Before the user selects from NumberPicker, we want the app to set the choices that the user will see in NumberPicker to the items in phoneNumbers. Then, after the user selects a number, we want NumberPicker to disappear, replaced by PlacePicker. Here are the event handlers that we need for step 1.

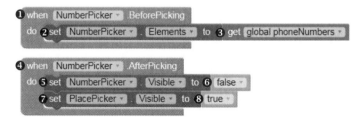

In the Blocks pane, click **NumberPicker** and drag the **whenNumberPicker.Before Picking** event handler block ❶ to the Viewer. Then, click **NumberPicker** again and drag its **setNumberPicker.Elementsto** block ❷ next to the word *do*. Next, in the Blocks pane, click the **Variables** blocks drawer and drag the **get global**

phoneNumbers block ❸ to the right of the setNumberPicker.Elementsto block. These three blocks create the BeforePicking event handler, which sets the telephone number choices that the user will see in NumberPicker to the telephone numbers in the phoneNumbers variable. Next we'll create the handler for the AfterPicking event.

Handling NumberPicker's AfterPicking Event

To replace NumberPicker with PlacePicker after the user chooses a telephone number in NumberPicker, drag the whenNumberPicker.AfterPicking event handler block ❹ to the Viewer. Then, drag the setNumberPicker.Visibleto block ❺ next to the word do. Next, click the **Logic** blocks drawer and snap the **false** block ❻ to the right of the setNumberPicker.Visibleto block. Then, click **PlacePicker** in the Blocks pane and place the setPlacePicker.Visibleto block ❼ under the setNumberPicker.Visibleto block. Finally, click the **Logic** blocks drawer again and snap the **true** block ❽ to the right of the setPlace Picker.Visibleto block.

To see how the blocks for step 1 work, live-test with a phone, as outlined in "Live-Testing Your Apps" on page xxii. Once you click **Connect ▸ AI Companion** in the top menu bar and scan the QR code with your phone's AI2 Companion app, your "Beat the Bus" app should open on your phone. As long as the blocks are placed as shown, under the app's title image, you should see the Map with the Marker pointing to the latitude and longitude you set in the Properties pane in the Designer. You may also see a user icon located at your current location, if your location and the Marker point are both visible at the Map's zoom level.

Below the map, you should see NumberPicker. When you click it, the list of numbers in your phoneNumbers variable should appear, and, after you select one, NumberPicker should disappear and PlacePicker should appear. That's all that should happen for now. Leave the app open on your phone to keep live-testing.

STEP 2: CHOOSING THE DESTINATION

In step 2, to code the BeforePicking event for the PlacePicker, we'll need to create a new global list variable called places to store the location choices we'll display to the user. We'll also create two related global list variables: placeLats, which will hold the latitudes for the destination options, and placeLongs, which will hold the longitudes.

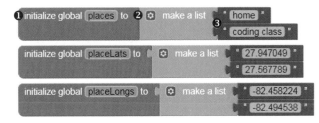

By creating these as global variables, we can use them in the Before Picking event handler for PlacePicker and in all other handlers in the app to access the name, latitude, and longitude for the user's selected location.

Creating Additional Global List Variables

For each variable, click the **Variables** block drawer and drag an initialize global name block ❶ to the Viewer, click name, and replace it with the variable name. Then drag a make a list block ❷ from the Lists drawer and snap it to the right side of the initialize global block. Finally, drag two empty string blocks ❸ from the Text drawer and snap them into the sockets of the make a list block, and enter data into each empty string block.

When you enter the data for placeLats and placeLongs, be sure that you place the coordinate values in the same positions as their corresponding item in the places variable. For example, the first value you enter for placeLats should be the latitude of the first value in your places variable, which is home in this list. The first value you enter for placeLongs should be the longitude of the first value in your places variable. Likewise, the second value you enter into placeLats must be the latitude for the second value in your places variable (coding class in this list), and the second value you enter into placeLongs should be the longitude for the second value in places.

Handling PlacePicker's BeforePicking Event

Now that we've created places, we have the data we need for PlacePicker. Before the user can select from PlacePicker, we want the app to set the choices that the user will see to the items in places. The PlacePicker BeforePicking event handler does this.

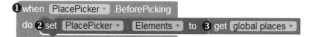

In the Blocks pane, click **PlacePicker** and drag the **whenPlacePicker .BeforePicking** event handler block ❶ to the Viewer. Then, click **PlacePicker** again and drag its **setPlacePicker.Elementsto** block ❷ next to the word do. Next, in the Blocks pane, click the **Variables** blocks drawer and drag the **get global places** block ❸ to the right of the **setPlacePicker.Elementsto** block. Those three blocks create the BeforePicking event handler, which sets the destination choices that the user will see in PlacePicker to the items in the places variable.

To see how these new blocks work, live-test again. If the blocks are placed as shown, once you choose a number and NumberPicker disappears, you should see PlacePicker. When you click it, the list of destinations in your places variable should appear, and, after you select one, nothing else should happen for now.

STEP 3: STARTING LOCATION TRACKING

Let's now program the PlacePicker AfterPicking event handler.

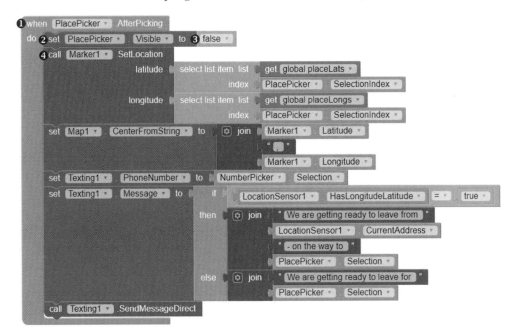

These blocks make PlacePicker disappear, move the Marker to the user's selected destination, and center the Map at that location. They also tell the app to text the user's starting location and destination to the selected phone number.

Making PlacePicker Disappear

In the Blocks pane, click **PlacePicker** and drag the **whenPlacePicker.AfterPicking** event handler block ❶ to the Viewer. Then, click **PlacePicker** again and drag its **setPlacePicker.Visibleto** block ❷ next to the word do. Next, in the Blocks pane, click the **Logic** blocks drawer, drag the **false** block ❸ to the Viewer, and snap it to the right of the **setPlacePicker.Visibleto** block. These blocks make PlacePicker disappear after the user chooses a destination.

Resetting the Map Marker Location

Now we'll add the blocks that set the location of the Marker to the destination selected by the user. In the Blocks pane, click **Marker1**, drag the **callMarker1 .SetLocation** method block ❹ to the Viewer, and snap it inside the **whenPlace Picker.AfterPicking** block under the setPlacePicker.Visibleto block. The Marker1 SetLocation method requires arguments for two parameters, latitude and longitude, and will set Marker1 on the Map to the location at the latitude and longitude values provided. We want those values to be the latitude and longitude for the destination chosen by the user in PlacePicker, which we get from the placeLats and placeLongs variables.

Setting the Map Marker Latitude and Longitude

Let's take a closer look at the blocks in the AfterPicking event handler that set the latitude and longitude parameter arguments for the Marker1 SetLocation method to the latitude and longitude for the user's destination.

Click the **Lists** blocks drawer, drag a `select list item` block ❶ to the Viewer, and snap it to the right of the `callMarker1.SetLocation` block, next to the word latitude. The select list item block requires arguments for list and index, and then, from the list provided, will select the item at the specified index position.

To provide those arguments, click the **Variables** blocks drawer and drag the `get global placeLats` block ❷ to the right of the `select list item` block next to the word list. Then click **PlacePicker** and drag the **PlacePicker.Selection Index** block ❸ next to the word index.

These blocks set the list argument for the select list item block to the placeLats variable and the index argument to PlacePicker.SelectionIndex (the index of the item the user selected in the related PlacePicker, which pulls its data from the places variable). For instance, if the user chose home as the destination, PlacePicker.SelectionIndex would equal 1, since home is the first element in PlacePicker from the places variable. If the user selected coding class, PlacePicker.SelectionIndex would equal 2. Altogether, these blocks set the latitude for Marker1 to the latitude listed in the placeLats variable at the PlacePicker.SelectionIndex position, which is the latitude for the user's selected destination.

To set the longitude for Marker1, copy the blocks at ❶, ❷, and ❸ and snap the copy to the right of the `callMarker1.SetLocation` block, next to the word longitude ❹. Then change the **list** argument in the duplicate blocks by clicking the drop-down arrow next to `get global placeLats` and selecting `get global placeLongs` ❺. These blocks set the longitude for Marker1 to the longitude value listed in the placeLongs variable at the PlacePicker.Selection Index position. The app will use this longitude and the latitude we just set to move Marker1 to the destination selected by the user.

Re-centering the Map

Now let's add the blocks to re-center the Map to the same location—the selected destination.

In the Blocks pane, click Map1, drag the setMap1.CenterFromStringto block ❶ to the Viewer, and snap it inside the whenPlacePicker.AfterPicking block under the callMarker1.SetLocation method block. Then, click the **Text** blocks drawer and drag a join block ❷ to the Viewer. Add an input to it by clicking the blue mutator icon to the left of the word join and dragging one additional string block to the join block in the dialog that opens.

Next, for the join block's top input, click Marker1 in the Blocks pane and drag in the **Marker1.Latitude** block ❸. For the join block's second input, drag in an empty string block ❹ from the **Text** blocks drawer, click the string block's text area, and enter , . Then, for the join block's bottom input, click Marker1 in the Blocks pane drawer again and drag in the **Marker1.Longitude** block ❺. After the user picks a destination, these blocks center the Map to Marker1's latitude and longitude, which are the coordinates for the destination the user selected in PlacePicker.

Texting the Selected Phone Number

The last action we need to program for step 3 is the texting. After the user selects a destination, we want the app to text the destination and the user's current location to the phone number the user chose. The following blocks in the AfterPicking event handler set the PhoneNumber for Texting1.

Click Texting1 in the Blocks pane, drag the setTexting1.PhoneNumberto block ❶ to the Viewer, and snap it inside the whenPlacePicker.AfterPicking block under the setMap1.CenterFromStringto block. Then click NumberPicker and drag the NumberPicker.Selection block ❷ to the Viewer to the right of the setTexting1.PhoneNumberto block. These blocks set the required phone number for Texting1 to the number selected by the user from the Number Picker drop-down list.

Deciding the Message to Send

Now we need to add the conditional blocks that set the message for Texting1 to send. That message will consist of one statement if the app has received the user's current longitude and latitude and another statement if the app has not yet received those coordinates. The next blocks in the AfterPicking event handler set the message.

Click **Texting1** again, drag the **setTexting1.Messageto** block ❶ to the Viewer, and snap it inside the **whenPlacePicker.AfterPicking** block under the **setTexting1.PhoneNumberto** block.

Then, click the **Control** blocks drawer, drag an **if then else** block with a left plug ❷ to the Viewer, and snap it to the right of the **setTexting1.Messageto** block. The left plug indicates that this block is a *return block*, which means it returns a value after evaluating a condition rather than only executing commands like the **if then** blocks we've used before.

Here, the **if then else** block directs the app to evaluate a condition and return one of two joined strings to use for **Texting1**'s **Message**—either the value from the blocks in the **then** socket or the value from the blocks in the **else** socket. To provide the condition we want the app to evaluate, click the **Logics** blocks drawer and drag an **=** comparison block ❸ to the Viewer next to the word **if**. To fill the **=** comparison operator block's left operand socket, click **LocationSensor1** in the Blocks pane and drag in its **LocationSensor1.HasLongitude Latitude** block ❹. Fill its right operand socket by clicking the **Logic** blocks drawer and dragging in the **true** block ❺. So far, these blocks set the test condition for the app to evaluate in order to determine the message to text, which is whether **LocationSensor1** has gotten the user's longitude and latitude.

Texting One Message If the App Knows the User's Longitude and Latitude

The next blocks tell the app the message to send if **LocationSensor1** has the user's longitude and latitude, which means the test condition evaluates to **true**.

Click the **Text** blocks drawer and drag a **join** block ❶ to the Viewer to the right of the word **then** in the **if then else** block. Add two more inputs to the **join** block by clicking the blue mutator icon to the left of the word **join** and dragging two additional string blocks in the dialog that opens. For the **join** block's top input, drag in an empty string block ❷ from the Text blocks drawer, click the string block's text area, and enter **We are getting ready to leave from** (with a space after the word **from**).

For the **join** block's second input, click **LocationSensor1** in the Blocks pane and drag in its **LocationSensor1.CurrentAddress** block ❸. For the **join** block's third input, drag in an empty string block ❹ from the Text blocks drawer, click the string block's text area, and enter **- on the way to** (with a space after the word to). Then, for the `join` block's bottom input, click **PlacePicker** in the Blocks pane drawer and drag in its **PlacePicker.Selection** block ❺. These blocks set the message for the app to send if the app's Location Sensor has received the user's longitude and latitude, which means the message will indicate the user's destination and also include the user's starting point—the current address.

Texting a Different Message If the App Doesn't Know the User's Longtitude and Latitude

Next, we'll add the blocks that dictate the message the app will text if LocationSensor1 hasn't yet received those coordinates, meaning the if then else test condition evaluates to false.

Click the **Text** blocks drawer and drag another `join` block ❶ to the Viewer to the right of the word else in the if then else block. For the join block's top input, drag in an empty string block ❷ from the Text blocks drawer, click the string block's text area, and enter **We are getting ready to leave for** (with a space after the word for).

For the join block's bottom input, click **PlacePicker** in the Blocks pane drawer and drag in another **PlacePicker.Selection** block ❸. These blocks set the message for the app to send to a statement that indicates the destination but does not include the user's current address, since the app's LocationSensor doesn't yet know it.

Sending the First Text Message

Finally, click **Texting1** again and drag its **callTexting1.SendMessageDirect** block into the **whenPlacePicker.AfterPicking** block under the setTexting1.Messageto block. This block sends the text message.

In plain English, the blocks for step 3 move the map Marker to the user's selected destination, re-center the map, and then direct the app to decide the message to text based on whether LocationSensor1 has retrieved the user's current latitude and longitude. If it has, the app should send a message to the selected telephone number that includes the user's current location and the destination. If not, the app sends a message that includes only the destination.

Use both your tester phone and the phone with the selected phone number to live-test these blocks. If you've placed the blocks exactly as shown, once you choose a number and NumberPicker disappears, you should see PlacePicker. When you click it, the list of destinations in your places variable should appear.

After you select a destination, `PlacePicker` should disappear and your `Map` and `Marker` should shift to the selected destination's latitude and longitude, if those properties are different from those you set in the Designer. Also, your phone should text a message to your selected number, as long as you have texting enabled on your phone. If anything's not working as planned, take the time to debug before moving on. Then close the AI2 Companion app on your phone.

STEP 4: TRACKING THE JOURNEY

Let's now program step 4 so that the app texts the user's location to the selected phone number periodically as the user travels the route, but only when the user is less than or equal to 5 miles away from the selected destination. This means that we'll need to use another conditional to check whether the user's current location is within 5 miles of the chosen location. Use the following `LocationChanged` event handler blocks.

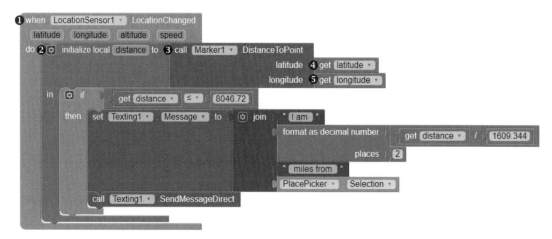

Click `LocationSensor1` in the Blocks pane and drag the `whenLocationSensor1 .LocationChanged` block ❶ to the Viewer. You'll notice that this `LocationChanged` event handler provides arguments for four event parameters when the app's location changes: the new location's `latitude`, `longitude`, `altitude`, and `speed`. These parameters are local in scope, meaning we can use their values only in the event handler in which they're created, just like the local variable we'll create in the next section.

Creating a Local distance Variable

To simplify the code, we'll use the arguments provided for the `LocationChanged` event handler parameters to create a local variable, `distance`, to hold the value of the app's current distance from `Marker1`, which we've positioned on the `Map` at the user's selected destination. Since `distance` is a local variable, we can use it in this event handler only.

Click the **Variables** blocks drawer and drag the **initialize local name to** block ❷ to the right of the `whenLocationSensor1.LocationChanged` block next to the word do. Click **name** in the **initialize local name to** block and replace it with **distance**. Then click **Marker1** and snap its **callMarker1.DistanceToPoint** method block ❸ to the right of the **initialize local distance to** block. To fill the method block's required latitude and longitude parameter sockets, mouse over the **latitude** event parameter above the word do and drag the **get latitude** block ❹ into the method's latitude socket; then, mouse over the **longitude** event parameter above the word do and drag the **get longitude** block ❺ into the method's longitude socket.

Now that we've created distance, we can use it in this event handler wherever we need to refer to the app's current distance from the user's selected destination.

Next, let's direct the app to text information about the user's location only if distance is less than or equal to 5 miles.

Deciding Whether to Text Current Distance

To test whether distance is less than or equal to 5 miles, we need to set another test condition using the following blocks.

❶ ⚙ if ❷ ❸ get distance ≤ ❹ 8046.72

Click the **Control** blocks drawer and drag an **if then** block ❶ inside the **initialize local distance to** block under the **callMarker1.DistanceToPoint** block. Now click the **Math** blocks drawer, drag an = comparison operator block ❷ to the Viewer next to the word if, and turn = into a ≤ comparison block by clicking the drop-down arrow to the right of the equal sign and selecting the less than or equal sign.

To fill the ≤ comparison block's left operand socket, mouse over the local **distance** variable and drag in its **get distance** block ❸. To fill the ≤ comparison block's right operand socket, click the **Math** blocks drawer, drag in a **8046.72** number block ❹ for the metric equivalent to 5 miles. This code sets the test condition that determines whether the user is less than or equal to 5 miles away from the selected destination. The condition must evaluate to true in order for the app to send a text message in this step.

Setting and Sending Text Messages Along the Route

The next blocks set the text of the message for the app to send if our test condition is met—that is, the user is less than or equal to 5 miles from the selected destination.

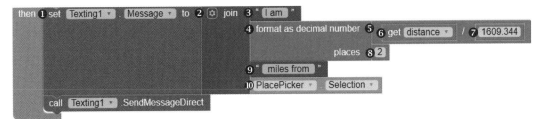

Click **Texting1** and drag its `setTexting1.Messageto` block ❶ into the `if then` block next to the word then. Now, click the **Text** blocks drawer and drag a `join` block ❷ to the Viewer to the right of the `setTexting1.Messageto` block. Add two more inputs to it by clicking the blue mutator icon to the left of the word `join` and dragging two additional string blocks to the `join` block in the dialog that opens. For the `join` block's top input, drag in an empty string block ❸ from the Text blocks drawer, click the string block's text area, and enter **I am** (including a space after the word am).

For the `join` block's second input, drag in a `format as decimal` block ❹ from the Math drawer and fill its `number` socket by dragging in a division operation block ❺, also from the Math drawer. Mouse over `distance` to drag its `get distance` block ❻ into the division block's left operand socket. Then, drag a `1609.344` number block ❼ into the division block's right operand socket for the metric equivalent to 1 mile. Finally, fill the `format as decimal number` block's `places` socket with a `2` number block ❽ from the Math drawer. This second `join` block input computes the user's distance in miles, presents it as a decimal number with two decimal places, and adds it to the text message the app will send. Since App Inventor calculates distances in meters, we're converting it to miles for users in the US.

Next, fill the `join` block's third input with another empty string block ❾ from the Text blocks drawer, click the string block's text area, and enter **miles from** (including a space before the word `miles` and after the word `from`). Then, for the `join` block's bottom input, click **PlacePicker** in the Blocks pane and drag in its `PlacePicker.Selection` block ❿.

Finally, click **Texting1** again, and drag its `callTexting1.SendMessageDirect` block into the `if then` block under the `setTexting1.Messageto` block to send the message. These blocks set the text message to "I am [*x.xx*] miles from [*selected destination*]" and send it to the phone number the user selected.

Now, altogether, the blocks for step 4 tell the app to text the user's location to the selected phone number periodically as the user is en route, but only if the user is less than or equal to 5 miles away from the destination.

Now we can test the completed app! To do so, you need to build the app and install it so you can run it as you are moving along the route. For testing, you can install the app on your phone quickly after creating a temporary QR code that you can scan with your phone's AI2 Companion app.

WARNING *Please do not try to test this app while you are driving. If you test in a vehicle, have someone else drive so that you can safely operate and monitor the two phones. You also can test the app while walking.*

With the app project open on your computer screen, click **Build ▸ App (provide QR code for .apk)** in the top menu bar, as shown in Figure 5-2.

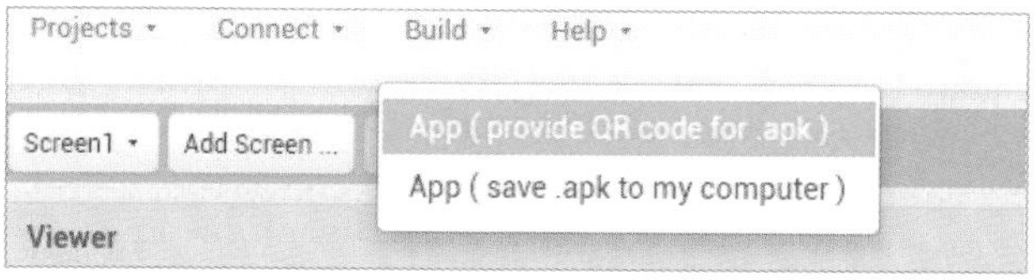

Figure 5-2: The Build menu link that creates the QR code you can scan to install the app

You should see a progress bar showing that the QR code is being generated. Once the code is ready, you should see a small window containing a temporary QR code link to your app's source file. Scan the code with the AI2 Companion app on your phone and click **OK** to close the small window.

Note that since "Beat the Bus" is a nonmarket app, you'll first need to make sure your phone's settings allow installation of applications from "Unknown Sources." When you install the app, be sure to allow it to receive text messages and access approximate location, precise location, and extra location provider commands.

Once you've installed and opened the app, you should see the Map with the Marker pointing to the latitude and longitude you set in the Properties pane in the Designer. You may also see a user icon located at your current location, if your location and the Marker are both visible at the Map's zoom level.

You should also see NumberPicker, which should disappear after you click it and choose a number. Then you should see PlacePicker, which will disappear after you click it and select a destination. Next, your Map and Marker should shift to the selected destination's latitude and longitude, if those properties are different from those you set in the Designer. The app should also text a message to your selected number.

As you travel along, you should see the user icon move occasionally on the Map, and when you get within 5 miles of your destination, the app should text periodic messages to your selected number until you close the app.

You may notice that, although you've set LocationSensor1 to check for location changes every second, the app may not send a message each second. According to the App Inventor documentation, the LocationSensor component receives location updates only when the location of the phone actually changes, and the chosen time interval is not guaranteed. Nonetheless, your selected phone number should receive several messages between the time you reach the 5-mile mark and arrive at the destination. Note that, if you feel the messages are coming in too frequently, you can increase the LocationSensor's TimeInterval in the Designer.

If anything's not working as planned, take time to debug. If you make any changes, you'll need to reinstall the app to test them. If you placed the blocks correctly, the app should work as expected. You've successfully created the "Beat the Bus" tracking app!

SUMMARY

In this chapter, you built the "Beat the Bus" app, which uses App Inventor's exciting LocationSensor, Texting, and Maps components along with ListPickers to allow others to track your location, without using the power-draining location services on their phone. While building this app, you learned how to create list variables that can hold multiple values and local variables that can make your code more efficient. You also continued to use the programming selection control structure to direct the app's flow.

In Chapter 6, you'll learn how to use App Inventor's built-in Control and Logic blocks to control an app's flow using the repetitive program structure. We'll use more lists with that structure as we create the "Tic Tac Toe" game app, which allows users to play the classic two-person game and keeps track of the players' turns.

ON YOUR OWN

Save new versions of "Beat the Bus" as you modify and extend it for the following exercises. You can find solutions online at *https://nostarch.com/ programwithappinventor/.*

1. Change the app so that it also provides the destination street address in all text messages sent and retrieves that information from a list variable.

2. Change the app so that it requires users to enter the destination instead of choosing it from the preset list.

3. Extend the app for Android devices so that the app responds to specific text messages from the selected number by texting the user's current location at that time. How will your algorithm change? What components and blocks will you need to add and modify?

6

TIC TAC TOE: USING LOOPS TO CREATE A GAME

Programmers often write code that repeats a set of actions. For instance, they may want to display a password request each time a user enters the wrong password or search for the same phrase in thousands of files. To avoid copying and pasting the same lines of code over and over again, we use the repetition programming control structure or *loops*, which are code segments that repeat a task a specific number of times or until a certain condition is met. It's easier to review and correct errors in a loop than multiple lines of repeated code.

USING LOOPS IN APP INVENTOR

In App Inventor, you can control the flow of apps with three types of loops: a for each number loop that performs an action for each number in a given range, a while loop that repeats an action until a tested Boolean condition returns false, and a for each item loop that performs the same action on each item in a list. Next we'll take a closer look at each of these.

THE FOR EACH NUMBER LOOP

We often use a for each number loop to calculate a series of numbers. The following event handler shows how a for each number loop calculates the sum of certain numbers in a given range.

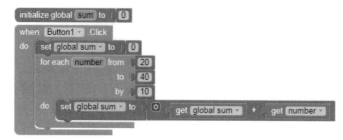

First, we initialize the global sum variable that holds the value to compute. Then we program the event handler to reset sum to 0 at each button click. This ensures that the app calculates the correct sum by erasing values from a previous click.

To create the for each number loop, we use a *counter variable* named number to count and control the number of times the app executes the code. Then we include a range of values, starting with the number in from and ending with the number in to. Inside the by socket, we insert a number that determines how much to increment the number variable between the loops. When you click the button, the app loops through and executes the blocks in the for each block's do socket for each number in the range from 20 to 40, while increasing the value of number by 10 between each loop.

This means that, in the first loop, the code sets the value of the global sum variable to its current value (0) plus the value of the number variable (20), which equals 20, and then increments the number variable to 30. In the second loop, it sets sum to 20 + 30, or 50, and then increments number to 40. In the final loop, it sets sum to 50 + 40, or 90, which is the sum we set out to compute. You can use this kind of code any time you want to compute the sum of numbers in a specific range incremented between each loop by a consistent number.

THE WHILE LOOP

A while loop repeats an action until a Boolean condition evaluates to false. The following event handler uses a while loop that chooses three random winners from 30 raffle tickets.

Before coding the event handler, we need to create an empty list variable named raffleWinners to hold the three randomly selected ticket numbers. We'll use raffleWinners along with a relational operator (<) to set the while loop's test condition. Before each loop, we test whether the length of, or number of items in, the raffleWinners list is less than 3.

We test for this condition because we want to execute the code in the do socket only three times to pick the three random raffle winners. As long as the length of raffleWinners is less than 3, we execute the code in the while loop's do socket. Once the length of raffleWinners reaches a value of 3, the loop ends, and we immediately execute the blocks beneath the loop.

When you click the button, while the test condition is true, the blocks in the do socket loop through and add a random integer between 1 and 30 to the raffleWinners list variable.

This means that, in the first loop when the length of raffleWinners is 0, the app adds one random number to raffleWinners, which then increases its length to 1. In the second loop, since the length of raffleWinners is still less than 3, the app repeats this action, adding another random number to raffleWinners, which then increases its length to 2. In the third loop, since the length of raffleWinners is still less than 3, the app adds another random number to raffleWinners, which then increases its length to 3, after which the test condition fails and the loop ends.

After the loop, the app displays raffleWinners, the list of randomly selected raffle winning numbers, in a label.

THE FOR EACH ITEM LOOP

We can use the for each item loop to repeat actions for each item in a list. Here is a for each item loop that we could add to "Beat the Bus" from Chapter 5 so that users can provide their destination to all numbers in the list of phone numbers, instead of just one.

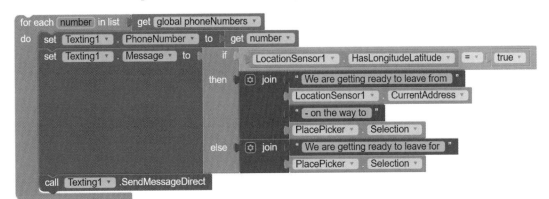

These blocks loop through the list of telephone numbers and text the location information to each number.

BUILDING THE "TIC TAC TOE" APP

In this chapter, you'll use two `for each item` loops and work more with the `Canvas` and `ImageSprite` components to create a "Tic Tac Toe" game app, where two players take turns touching empty squares on the `Canvas`. The first player who touches three empty squares in a row, either horizontally, vertically, or diagonally, wins the game. The app keeps track of the players' turns and lets them reset the game to play again at any time.

Log in to App Inventor and create a project called `TicTacToe`.

DECOMPOSING "TIC TAC TOE"

We can decompose the game activity into three steps:

1. When the app opens, display a nine-square Tic Tac Toe board and a reset button. Place an `ImageSprite` on top of each square on the board, which will display the X or O.

2. When the first player touches an empty square on the board, display an X on that square. Then display a message that shows it's the other player's turn (O), and so on. When the last square gets filled, tell the players that the game is over!

3. When a player hits the reset button, remove all Xs and Os from the board along with all notices, and the app is ready for a new game.

   You'll need the following components:

- **Button** for a player to click to reset the game board
- **Canvas** with a game board background for players to touch to play the game and on which to display X and O `ImageSprites`
- Global variable (3) to store `ImageSprites` and keep track of the current game player and number of plays
- **ImageSprite** (9) to display X and O graphics on empty game board squares touched by players
- **Label** to display game notices
- Local variable to store the value of the touched `ImageSprite`

LAYING OUT "TIC TAC TOE" IN THE DESIGNER

Now let's lay out the app in the Designer starting with `Screen1`. In the Properties pane for `Screen1`, set its **AlignHorizontal** property to **Center: 3** so that all components we place will sit in the middle of the screen toward the top.

Then, to keep the screen's orientation from changing when the device rotates, set **ScreenOrientation** to **Portrait**. Finally, remove the checkmark under **TitleVisible** to keep the title of the screen from showing when the app runs on a device.

Next, drag the other necessary components from their drawers in the Palette pane onto the Viewer pane and adjust their properties. Drag in the Label from the User Interface drawer that will show the player's turns. In the Properties pane, make its font bold and italic by clicking the checkboxes under FontBold and FontItalic, and change the FontSize to 30. Also, change the **Width** to **Fill parent**, remove the text under **Text**, set the TextAlignment to **center: 1**, and select **Magenta** as the TextColor.

To create the reset button, place a Button from the User Interface drawer under the Label. In the Properties pane, change the Button's BackgroundColor also to **Magenta**, make its FontBold, and increase FontSize to 20. Also, change the **Width** to **Fill parent**, and set the **Text** to Reset Game and TextColor to **White**.

Now drag in a Canvas from the Drawing and Animation drawer. Upload a background image of the Tic Tac Toe board by clicking the text box under BackgroundImage in the Properties pane and following the image upload instructions in "Uploading a Picture" on page 27. Also, set the Canvas's **Height** and **Width** to 100 percent.

Drag onto the Canvas nine ImageSprites that will show the Xs and Os when the players touch the screen. In the Properties pane, set the X and Y properties to 0 for those we want positioned in the three top and three left squares of the Tic Tac Toe board. We can do this now since we know that the 0 coordinate values for these five ImageSprites won't change no matter what size device a player uses. Set the 0 coordinates shown in Table 6-1 as you drag the ImageSprites to the Canvas.

Table 6-1: Initial X and Y Property Values for Top and Left ImageSprites

ImageSprite	X property	Y property
ImageSprite1	0	0
ImageSprite2		0
ImageSprite3		0
ImageSprite4	0	
ImageSprite7	0	

We won't set the other coordinates now, since their positions will change depending on screen size. Instead, we'll program the app to set them as a fraction of the Canvas size when the app opens.

Finally, upload the X and O images by clicking **Upload File...** in the Media pane and following the image upload instructions.

Now all components should show in the Viewer, except for Label1, which doesn't contain any text to start. You also should see the list of all app components in the Components pane and the names of the uploaded images in the Media pane, and Screen1 should look like Figure 6-1 (although, depending on where you dragged the ImageSprites, they may not appear as neatly ordered).

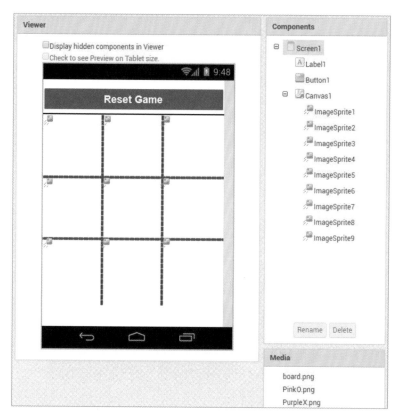

Figure 6-1: The Viewer, Component, and Media panes after you've laid out "Tic Tac Toe"

Now you're ready to program the app in the Blocks Editor.

PROGRAMMING "TIC TAC TOE"

For "Tic Tac Toe," we'll program three event handlers. The first directs the app's action when the screen initializes, and the others respond to the Canvas TouchDown and Button Click events. We'll also create four variables to store important information, including a global list, and we'll use several if then and two for each item loop blocks to direct the flow of the app's action. Click the **Blocks** button to switch to the Blocks Editor, and let's begin programming step 1.

STEP 1: SETTING UP THE TIC TAC TOE GAME BOARD

When the screen opens, we want players to see the nine-square Tic Tac Toe board and the reset button that we placed and styled in the Designer. At the same time, we want the app to respond to the Screen1 Initialize event by setting the width and height of the ImageSprites and positioning one on each square of the game board. The `Screen1 Initialize` event handler programs the behind-the scenes action in step 1.

```
when Screen1 . Initialize
do  for each sprite in list  get global squares
    do  set ImageSprite. Width of component  get sprite  to  ⚙  Canvas1 . Width  ×  0.333
        set ImageSprite. Height of component  get sprite  to  ⚙  Canvas1 . Height  ×  0.25

    set ImageSprite2 . X  to  ⚙  Canvas1 . Width  ×  0.333
    set ImageSprite5 . X  to  ⚙  Canvas1 . Width  ×  0.333
    set ImageSprite8 . X  to  ⚙  Canvas1 . Width  ×  0.333
    set ImageSprite3 . X  to  ⚙  Canvas1 . Width  ×  0.666
    set ImageSprite6 . X  to  ⚙  Canvas1 . Width  ×  0.666
    set ImageSprite9 . X  to  ⚙  Canvas1 . Width  ×  0.666
    set ImageSprite4 . Y  to  ⚙  Canvas1 . Height  ×  0.25
    set ImageSprite5 . Y  to  ⚙  Canvas1 . Height  ×  0.25
    set ImageSprite6 . Y  to  ⚙  Canvas1 . Height  ×  0.25
    set ImageSprite7 . Y  to  ⚙  Canvas1 . Height  ×  0.5
    set ImageSprite8 . Y  to  ⚙  Canvas1 . Height  ×  0.5
    set ImageSprite9 . Y  to  ⚙  Canvas1 . Height  ×  0.5
```

Creating the Global squares List Variable

Before you code the event handler, you should create the global `squares` list variable to store the nine `ImageSprites` that will display Xs and Os. Since the variable is global, we can use it in all event handlers to refer to the nine `ImageSprites`. The following blocks declare and initialize the variable.

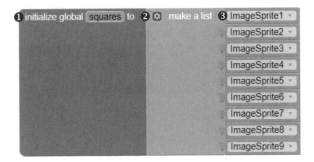

To place the blocks, click the **Variables** block drawer and drag an `initialize global name` block ❶ to the Viewer, click **name**, and replace it with `squares`. Then drag a `make a list` block ❷ from the Lists drawer, snap it to the right side of the **initialize global squares** block, and provide a socket for each of the nine ImageSprites by adding seven more sockets to the **make a list** block's two default inputs. To add the seven sockets, click the blue mutator icon to the left of the words make a list, and drag the additional item blocks to the **list** block in the dialog that opens.

Then, in the Blocks pane, click each **ImageSprite** in order from 1 to 9, and drag its **ImageSprite** getter block ❸ into its socket in the **make a list** block. We've now set the list items in squares to the nine ImageSprites.

Looping Through ImageSprites in the squares List

Now that you have the squares list, start the event handler by clicking **Screen1** in the Blocks pane and dragging the whenScreen1.Initialize block to the Viewer. Then add a for each item block that loops through the list of ImageSprites and sets their width and height to match those of the game board squares.

Because the list contains a series of components, you'll use the generic Any component blocks to access and manipulate the properties of all ImageSprites in the loop without needing to specifically name each ImageSprite. You can find Any component blocks in the Blocks Editor window beneath the component-specific blocks in the Blocks pane, as shown in Figure 6-2.

Any app you create in App Inventor will include Any component blocks for each component added.

Figure 6-2: The "Tic Tac Toe" app's Any component blocks

Setting the Width and Height of Each ImageSprite in the squares List

By setting the Canvas height and width to 100 percent of the screen size in the Designer earlier, we made sure both the Canvas and the squares in the Canvas's game board background image will change proportionately whenever the device's size changes. Likewise, by setting the width and height to a fraction of the Canvas size in the for each item loop, we make sure that the width and height of each ImageSprite will change proportionately with the size of the game board.

To place the loop, drag a **for each item** block ❶ from the Control blocks drawer to the Viewer and snap it inside the **whenScreen1.Initialize** block next to the word do. On the for each item block, you'll notice the counter variable item, which represents each item in the list. Click the word **item** and enter a more meaningful name, **sprite**.

Then click the **Variables** blocks drawer, drag in a `get global squares` block ❷, and snap it to the right side of the `for each sprite` block next to the word `list`. So far, for each sprite in the global squares list variable, these blocks will cause the app to perform some action. Now let's have the app set the width and height for each sprite.

To set the `ImageSprites`' width to one-third the `Canvas` width, click the plus sign to the left of **Any component** in the Blocks pane, click the **Any ImageSprite** drawer, and drag a `setImageSprite.Width` block ❸ inside the `for each sprite` block next to the word `do`. This generic setter block requires us to identify the `ImageSprite` we're changing before providing the size we're changing it to. To identify these values, fill its `of component` socket by mousing over the **sprite** counter variable and dragging in a `get sprite` block ❹. Fill its `to` socket with a multiplication operator block ❺ filled on the left with a `Canvas1.Width` block ❻ from the Canvas1 drawer and on the right with a `0.333` number block ❼.

To set the height of each sprite to one-fourth the `Canvas` height, make a copy of the `setImageSprite.Width` blocks and snap the duplicate blocks inside the `for each sprite` block under the `setImageSprite.Width` block. In the duplicate blocks, click the drop-down arrows in both the `setImageSprite.Width` block ❽ and the `Canvas1.Width` block ❾ and select `Height`. Replace `0.333` in the number block with `0.25` ❿.

Positioning ImageSprites on the Canvas

Finally, we need to place each `ImageSprite` on the game board so that it lines up with the top-left point of its corresponding square, as shown in Figure 6-3.

We want the x- and y-coordinates of `ImageSprite1` to be the top-left corner of square 1 (which is where X = 0 and Y = 0), the x- and y-coordinates of `ImageSprite2` to be the top-left corner of square 2, and so on.

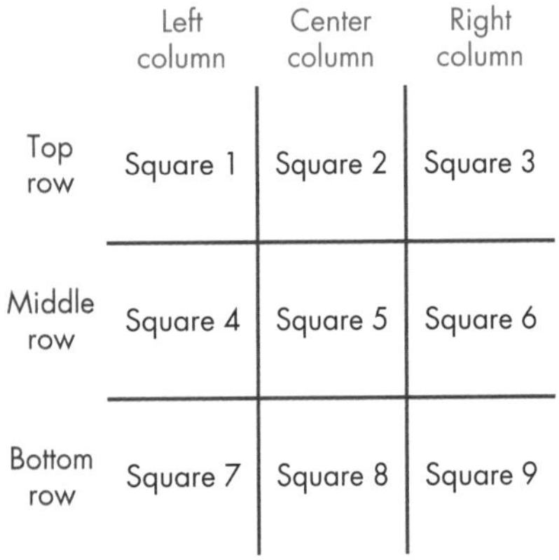

In the Designer, we already set the X property of `ImageSprite1`, `ImageSprite4`, and `ImageSprite7` to 0, because we want them in the left column of the `Canvas` over squares 1, 4, and 7. We also set the Y property of `ImageSprite1`, `ImageSprite2`, and `ImageSprite3` to 0, because we want them at the top row of the `Canvas` covering squares, 1, 2, and 3.

The following blocks set the `ImageSprites`' remaining x- and y-coordinates.

	Left column	Center column	Right column
Top row	Square 1	Square 2	Square 3
Middle row	Square 4	Square 5	Square 6
Bottom row	Square 7	Square 8	Square 9

Figure 6-3: Tic Tac Toe's Canvas background image with labeled squares, columns, and rows

First, set the x-coordinate of ImageSprite2 to the farthest point at the left of square 2 on the game board in Figure 6-3, which is the Canvas width multiplied by 0.333. Place the **setImageSprite2.Xto** block ❶ inside the **when Screen1.Initialize** block under the **for each** block. Then attach a multiplication operator block to the **setImageSprite2.Xto** block ❷, and fill it with a **Canvas1.Width** block ❸ and a **0.333** number block ❹. Now make 11 duplicates of the **setImageSprite2.Xto** block and make the changes shown in the code.

These duplicate blocks set the x-coordinates of ImageSprite5 and Image Sprite8 to the Canvas width multiplied by 0.333, which is the farthest point at the left of the Tic Tac Toe board's center column on the board in Figure 6-3, and the x-coordinates of ImageSprite3, ImageSprite6, and ImageSprite9 to the Canvas width multiplied by 0.666, which is the farthest point at the left of the Tic Tac Toe board's right column on the board in Figure 6-3.

They also set the y-coordinates of ImageSprite4, ImageSprite5, and Image Sprite6 to the Canvas height multiplied by 0.25, which is the very top point of the Tic Tac Toe board's middle row on the board in Figure 6-3, and the y-coordinates of ImageSprite7, ImageSprite8, and ImageSprite9 to the Canvas height multiplied by 0.5, which is the very top point of the Tic Tac Toe board's bottom row on the board in Figure 6-3.

Now, with all of our step 1 blocks, we've programmed the app to respond to the Screen1 Initialize event by placing on each Tic Tac Toe board square an ImageSprite that shows no image, has the same width and height as the square, and is positioned with its top-left point at the top-left point of the square.

Let's live-test now with a device, as outlined in "Live-Testing Your Apps" on page xxii. Click **Connect ▸ AI Companion** in the top menu bar and scan the QR code with your device's AI2 Companion app. Your "Tic Tac Toe" app should open on your device, and you should see the game board and the reset button.

The app should have the invisible ImageSprites on the Canvas, and you should see their width, height, and position after we program the Canvas

TouchDown event handler in step 2. Leave the app open on your device to keep live-testing.

STEP 2: RESPONDING TO PLAYER MOVES

In step 2, we'll program the app to respond each time a player touches an empty square on the Canvas Tic Tac Toe game board. That's when we want the app to perform several actions:

1. Determine which ImageSprite the player touched using the x- and y-coordinates of the place touched and a series of nested conditionals.
2. Determine whether player X or O touched the Canvas using the data stored in a global variable called player.
3. Display a corresponding X or O graphic in the touched ImageSprite.
4. Change the value of player to the opposite player.
5. Keep track of the number of plays in a global variable called play. (There's a total of nine possible plays in Tic Tac Toe.)
6. Display a notice indicating who plays next, until the value of play equals 9, when the notice should then read, "Game's over!"

We'll start the code for step 2 by creating the two global variables: player and play.

Creating Global Variables

The first global variable, player, stores the value of the current player, either X or O. The second global variable, play, keeps track of the number of plays made in the game. Since both player and play are global, we can use them throughout the code in all event handlers. The following blocks initialize the player and play variables.

❶ initialize global `player` to **❷** " X "
initialize global `play` to **❸** 0

To create each variable, click the **Variables** block drawer and drag an `initialize global name` block **❶** to the Viewer, click `name`, and replace it with the variable name. Then, for `player`, since it will hold string data, drag an empty string block **❷** from the Text drawer, snap it to the right side of the `initialize global player` block, and give it the initial value for the first player, player X, by entering X in the empty string block.

Initialize `play`, which will hold numeric data representing the number of plays in the game, to the value of 0 by dragging a 0 number block **❸** from the Math drawer and snapping it to the right side of the `initialize global play` block.

Handling the Canvas TouchDown Event

Now that we've created player and play, we have the data we'll need for the step 2 Canvas TouchDown event handler.

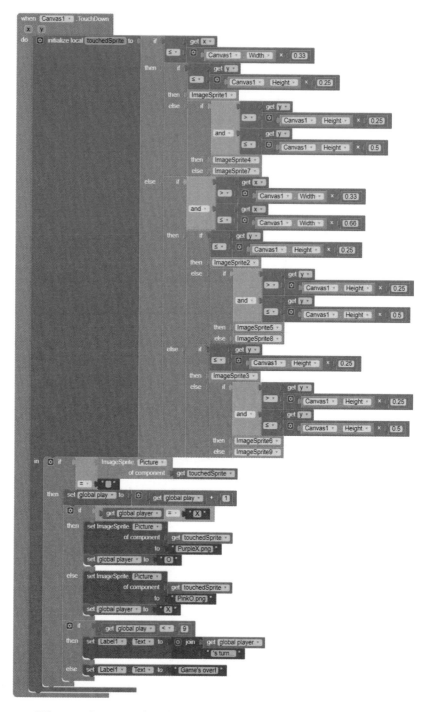

When a player touches down on an empty square on the Canvas game board, we need the app to place the X image on the touched spot if player X touched it, and the O image if player O touched it. To do this, the app first needs to know which square on the board and corresponding ImageSprite the player touched. Fortunately, the Canvas TouchDown event handler includes two event parameters, x and y, that hold the x- and y-coordinates of the place touched. The app will use these coordinates to determine the game board square and ImageSprite touched.

We start by using the x and y event parameters to set the value of a local variable called touchedSprite to the ImageSprite that was touched. We do this by first testing a series of conditions to determine whether the player touched a square and its corresponding ImageSprite in the left column of the game board. If the player didn't touch the left column, we test whether the player touched the center column, and then, if necessary, the right column.

Once we know the value of touchedSprite—the ImageSprite that the player touched—then the app can determine whether that ImageSprite is already displaying an image, and, if not, display either the X or O image, depending on which player touched the board.

Creating the Local touchedSprite Variable

The following blocks from the Canvas TouchDown event handler determine which ImageSprite the player touched and then set it as the value of touchedSprite.

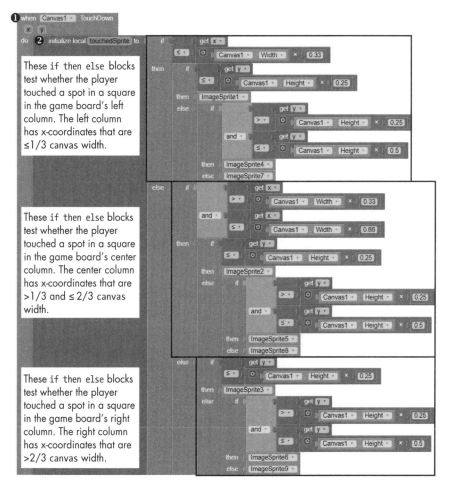

First, click `Canvas1` in the Blocks pane and drag its `whenCanvas1.TouchDown` event handler block ❶ to the Viewer. Then, create the local variable touched Sprite by clicking the **Variables** blocks drawer, dragging the first `initialize local name to in` block inside the `whenCanvas1.TouchDown block` ❷ next to the word do, and changing `name` to `touchedSprite`.

This block sets the initial value of the local variable `touchedSprite` to the value of the blocks we snap into the to socket. Since `touchedSprite`'s scope is local, we can use it only in this `TouchDown` event handler in the blocks we snap next to the word in later in this step.

Testing Whether touchedSprite Is in the Left Column of the Game Board

Now we'll add the blocks that set the value for `touchedSprite`, which is the `ImageSprite` that displays on top of the square a player has touched. To figure out which spot on the board was touched, the app first tests whether the spot is located in the left column of the board in Figure 6-3, which would mean the x parameter is less than or equal to the `Canvas` width multiplied by 0.33.

To place the blocks, click the **Control** blocks drawer and place an `if then else` block ❶ in the `initialize local touchedSprite to in` block's to socket.

Fill the `if then else` block's `if` socket with a ≤ comparison operator block ❷ from the Math drawer. Fill the ≤ operator block's first socket by mousing over the `whenCanvas1.TouchDown` block's x parameter and dragging in the `get x` block ❸. Fill the ≤ operator block's second socket with a multiplication operator block ❹. Now fill the multiplication block by clicking `Canvas1` in the Blocks pane and dragging its `Canvas1.Width` block ❺ into the left operand socket and snapping a `0.33` number block ❻ into the right operand socket.

These blocks check whether the sprite the player touched is in the game board's left column, or `ImageSprite1`, `ImageSprite4`, or `ImageSprite7`.

Testing Whether touchedSprite Is in the Top Row of the Left Column

If the first condition is true, we want the app to test the second condition, which is whether the y parameter of the touched spot is less than or equal to the `Canvas1` height multiplied by 0.25, or located in the top row of the game board. The following blocks test this second condition.

Click the **Control** blocks drawer, drag in the second `if then else` block ❶, and snap it to the right of the first `if then else` block next to the word then.

Next, fill the second `if then else` block's `if` socket with a ≤ operator block ❷. Then fill the ≤ operator block's first socket by mousing over the `whenCanvas1.TouchDown` block's y parameter and dragging in the `get y` block ❸, and fill its second socket with a multiplication operator block ❹. Fill the multiplication block by dragging the **Canvas1.Height** block ❺ into the left socket and a **0.25** number block ❻ into the right socket. Finally, fill the second **if then else** block's **then** socket by clicking **ImageSprite1** in the Blocks pane and dragging in its **ImageSprite1** getter block ❼.

Since we already know the player touched ImageSprite1, ImageSprite4, or ImageSprite7, we test whether the touched sprite is in the top row of the game board as shown in Figure 6-3. If so, then we know that touchedSprite is ImageSprite1 because it's the only ImageSprite that meets both the first and second test conditions.

Testing Whether touchedSprite Is in the Middle Row of the Left Column

If the touchedSprite is in the left column of the game board but not in the top row, then the second condition would fail. In that case, we use the if then else block's else socket to test whether the sprite is in the middle row of the game board.

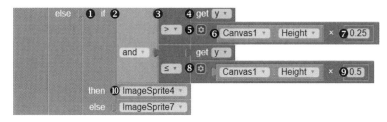

Place another **if then else** block ❶ inside the second **if then else** block next to the word else. Next, fill the third **if then else** block's **if** socket with an **and** logic operator block ❷ from the Logic drawer. The and logic block requires that both of its operands evaluate to true in order for the tested condition to be true.

The first and operand determines whether the y-coordinate of the touched spot is greater than the Canvas1 height multiplied by 0.25. Fill the **and** logic block's first socket with a **>** operator block ❸, and fill the first socket of the **>** operator block with the **get y** block ❹ and the second socket with a **×** operator block ❺ that multiplies Canvas1.Height ❻ on the left by 0.25 ❼ on the right.

Now add the second and operand, which determines whether the y-coordinate of the touched spot is less than or equal to the **Canvas1** height multiplied by 0.5. Fill the **and** logic block's second socket by snapping a copy of the blocks in its first socket, clicking the drop-down arrow next to the **>** operator to change it to a **≤** operator block ❽, and changing the **0.25** number block to **0.5** ❾.

Finally, add the blocks that tell the app the result if both of those conditions are true, which means that the touchedSprite is ImageSprite4. If either condition is not true, it means that touchedSprite is ImageSprite7. Fill the third **if then else** block's **then** socket by clicking **ImageSprite4** in the Blocks pane and dragging in its **ImageSprite4** getter block ❿ and its **else** socket by dragging in the **ImageSprite7** getter block.

Now, if touchedSprite meets the first condition by being in the left column of the game board but fails the second condition because it's not in the top row, the blocks test the third condition—whether the touched sprite is in the middle row of the game board. If so, then we know touchedSprite is Image Sprite4, because it's the only ImageSprite that meets both the first and third conditions by being in both the left column and the middle row of the game board.

But if the third condition is not true, or the touched sprite is not in the middle row of the game board, then we know that touchedSprite must be ImageSprite7, because it's the only remaining ImageSprite that meets the first condition—being located in the left column.

The rest of the blocks in the to socket of the initalize local touchedSprite block fill the else socket of the first if then else block and tell the app what to do if the touched sprite isn't in the left column of the game board. In that case, we test whether the touched sprite is in the center column of the game board and, if not, then test whether it's in the right column of the board.

To create the blocks to test whether the touched sprite is located in the center column, simply right-click the first **if then else** block and make a copy. Then snap the copy of the blocks into the first **if then else** block's **else** socket, and make the few changes indicated in the following section.

Testing Whether touchedSprite Is in the Center Column of the Game Board

To determine which ImageSprite is touchedSprite, the first three if then else blocks test whether the spot touched is in the left column of the board. Next, we need to handle cases where touchedSprite isn't in the left column at all by testing whether the spot touched on the game board is located in the center column. This would mean that the x parameter of the touched spot is greater than the Canvas width multiplied by 0.33 and less than or equal to the Canvas width multiplied by 0.66, and that touchedSprite is either ImageSprite2, ImageSprite5, or ImageSprite8.

These three if then else blocks test whether the touched spot is in the center column of the game board.

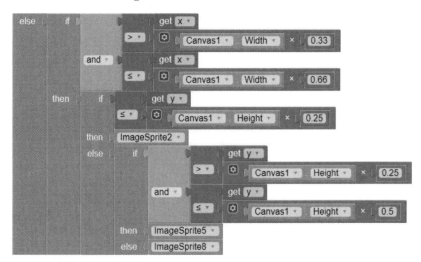

Make changes to the duplicate blocks that you snapped into the when Canvas1.TouchDown event handler in the first if then else block's else socket so they look exactly like those shown here.

These if then else blocks now represent the fourth condition we want the app to test—whether the touched sprite is in the center column of the game board—and tell the app what to do if the condition is met. If the touched sprite is in the game board's center column (meaning it's either ImageSprite2, ImageSprite5, or ImageSprite8), the blocks test the fifth condition, which is whether the touched sprite is in the top row of the game board. If this fifth condition is also true, then we know that touchedSprite is ImageSprite2.

But if the fifth condition is not true, the blocks test the sixth condition— whether the touched sprite is in the middle row of the game board. If so, then we know that touchedSprite is ImageSprite5, because it's the only Image Sprite that meets both the fourth and sixth conditions, meaning it's in both the center column and the middle row of the game board.

However, if the sixth condition is not true, meaning the touched sprite is not in the middle row of the game board, then touchedSprite has to be ImageSprite8, because it's the only remaining ImageSprite that meets the fourth condition—being located in the center column.

Determining Whether touchedSprite Is in the Right Column of the Game Board

So far, the first six if then else blocks test whether the spot touched on the game board is located in the left or center columns of the board. Now we need to add blocks to handle cases in which touchedSprite isn't in the left or center column but in the right column instead.

To determine whether the player touched ImageSprite3, ImageSprite6, or ImageSprite9, add the following blocks.

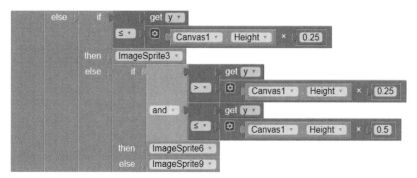

To place the blocks, copy the last two **if then else** blocks, snap the copy into the **whenCanvas1.TouchDown** event handler in the fourth **if then else** block's **else** socket, and make changes so the blocks are identical to those shown here.

These last two if then else blocks now pose the seventh condition we want to test: whether the touched sprite, which by default must be in the right column, is also in the top row of the game board. If it is, then touchedSprite is ImageSprite3.

But if the seventh condition is not true, the blocks test the eighth condition—whether the touched sprite is in the middle row of the game board. If so, then touchedSprite is ImageSprite6. However, if the eighth condition is not true, meaning the touched sprite is not in the middle row of the game board, then touchedSprite is ImageSprite9, because it's the only ImageSprite left.

Now that we've assigned a value to touchedSprite (the ImageSprite that was touched by the player), we'll use that value in the remainder of the whenCanvas1.TouchDown event handler. If the player touched an empty square on the board, we'll use the data stored in player to determine whether it was player X or O who touched the board, or played; display a corresponding X or O graphic in the touched ImageSprite; and then change the value of player to the opposite player.

Also, since there are a total of nine possible plays in Tic Tac Toe, we'll keep track of the number of plays in the play variable and display a notice indicating who plays next if the value of play is less than 9 and a notice that the game's over if play is 9 or greater. The following blocks from the when Canvas1.TouchDown event handler program these actions.

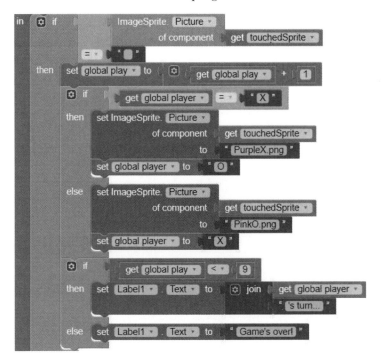

Testing Whether the Player Touched an Empty Square on the Game Board

Now we need to determine whether the player touched an empty square on the board. This would mean that the name of the touched ImageSprite's Picture is an empty string. The following blocks test that condition and, if the answer is yes, add 1 to the value of the variable that keeps track of the number of game plays.

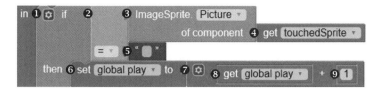

To add the blocks that test whether the name of the picture showing on touchedSprite is an empty string, drag an **if then** block ❶ from the Control drawer and snap it inside the **whenCanvas1.TouchDown** event handler next to the word in in the **initialize local touchedSprite to in** block. Then snap an = comparison operator block ❷ from the Logic drawer to the right of the word if.

Fill the first socket of the = block by clicking the **Any ImageSprite** drawer in the Blocks pane, dragging in an **ImageSprite.Picture** block ❸, and filling its socket with a **get touchedSprite** block ❹ from the Variables drawer. Fill the = operator block's second socket with an empty string block ❺ from the Text drawer.

If the player touched an empty square on the board, then we want the app to count the player's touch as a true game play and increment the value of the variable play by 1. To increment play by 1 in the code, snap the **set global play** block ❻ from the Variables drawer to the right of the word then and an addition operator block ❼ next to the **set global play** block. Then fill the addition operator block on the left with the **get global play** block ❽ and on the right with a **1** number block ❾.

Displaying the Correct Image on touchedSprite

Next we'll tell the app which image to display on touchedSprite based on the value of the player variable and then change the value of player to the opposite player.

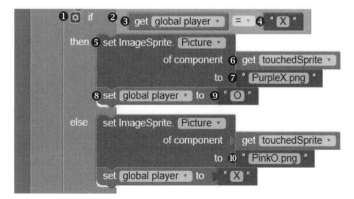

First we'll set up the test condition that determines if the value of the player variable equals X, which means that player X touched the Canvas. Place another **if then** block ❶ inside the **whenCanvas1.TouchDown** block under the **set global play** block. Click the **if then** block's blue mutator icon to add an **else** input to the block. Then, snap an = operator block ❷ from the

Logic drawer into the `if` socket, and fill its left side with a `get global player` block ❸ and its right side with a string block ❹ containing the letter X.

If the answer to the test condition is yes and the player variable equals X, we display X on `touchedSprite` and then change the value of `player` to 0. To do this, fill the `if then else` block's then socket by clicking the **Any ImageSprite** drawer in the Blocks pane, dragging in a `setImageSprite.Picture` block ❺, and filling its `of component` socket with a `get touched Sprite` block ❻ from the Variables drawer and its `to` socket with a string block ❼ from the Text drawer containing *PurpleX.png*, which is the name of the X graphic. Then, under the `setImageSprite.Picture` block, drag in a `set global player` block ❽ from the Variables drawer, and snap a string block ❾ from the Text drawer containing the letter 0 to the right of the word `to`.

If the answer to the test condition is no, meaning the `player` variable equals 0, we display O on `touchedSprite` and change the value of `player` to X. Code this by copying the blocks we just placed in the `if then else` block's `then` socket, snapping the copy into the `if then else` block's `else` socket, and making two changes. Change the name of the graphic in the `setImageSprite` `.Picture` block's bottom socket ❿ to `PinkO.png`. Then, change the 0 in the string block to the right of the `set global player` block to the letter X.

Altogether, the code tests whether the value of `player` is X, and, if so, direct the app to display the X graphic on `touchedSprite` and then change the value of `player` to 0. If the value of `player` is not X, they tell the app to display the O graphic on `touchedSprite` and then change the value of `player` to X.

Displaying Information on the Game's Label

The next blocks use a conditional to determine what to display on the app's `Label`, depending on the value of the variable `play`, which we've already programmed the app to increase each time a player touches an empty square on the board. If `play` is less than 9, then we want the `Label` to indicate who plays next. But, if `play` equals or exceeds 9, we want the `Label` to say the game is over.

To add the blocks, drag another `if then` block ❶ from the Control drawer and snap it inside the `whenCanvas1.TouchDown` block under the prior `if then else` block. Click the new `if then` block's blue mutator icon to add an `else` input. Then, to test whether `play` is less than 9, snap a `<` comparison operator block ❷ into the `if` socket and fill its left side with a `get global play` block ❸ from the Variables drawer and its right side with a `9` number block ❹

If the value of `play` is less than 9, the next blocks tell the app to indicate who plays next. Fill the `then` socket by clicking `Label1` in the Blocks pane, dragging in its `setLabel1.Textto` block ❺, and then snapping a `join` block ❻ from the Text drawer next to the word to. Fill the `join` block's top socket with the `get global player` block ❼ from the Variables drawer and the bottom socket with a string block ❽ from the Text drawer containing 's turn... to indicate which player's turn it is.

Finally, if play is not less than 9, the next blocks tell the app to say the game is over. Fill the `if then else` block's `else` socket by dragging in another `setLabel1.Textto` block ❾ and snapping a string block ❿ from the Text drawer containing Game's over! next to the word to.

Altogether, these blocks test whether the value of play, which holds the number of times a player has played in the game, is less than 9. If so, they direct the app to use the value of player to indicate whose turn it is to play next. However, if the value of play is 9 or more, they tell the app to indicate that the game is over.

Let's live-test now to see how the step 2 blocks work. As long as the blocks are placed correctly, once you open the app, you should see the game board and the reset button and be able to play the game.

When you touch an empty square, an X should appear in the square you touched, and the label above the reset button should display O's turn. Then, when the next empty square is touched, an O should appear in the touched square, and the label above the reset button should read X's turn, and so on, until the last empty square is touched and the label reads Game's over!. Since we haven't yet programmed the reset button, nothing should happen when you click it. We'll do that next.

STEP 3: PROGRAMMING THE RESET BUTTON

Let's now program step 3 of the app—coding the Button1 Click event handler. When a player hits the reset button, we want all X and O images on the game board and all text in the label to disappear, and we want to reset player and play to their original values. The Button1 Click event handler programs these actions.

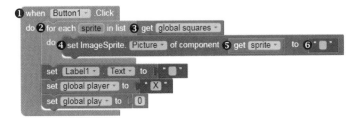

Clearing the Game Board

To make the X and O images disappear, we'll use another for each item and Any ImageSprite blocks to make the app loop though each ImageSprite and set the name of its Picture to an empty string so that it will no longer display an image.

Click **Button1** in the Blocks pane and drag its **whenButton1.Click** event handler block ❶ to the Viewer. Then drag in a **for each item** block ❷ from the Control drawer and snap it inside the **whenButton1.Click** block next to the word do, and then change **item** to **sprite**.

Next, drag a **get global squares** block ❸ from the Variables drawer and snap it to the right of the **for each sprite** block next to the word list. Then, next to the word do, drag in a **setImageSprite.Picture** block ❹ from the Any ImageSprite drawer, and fill its first socket by mousing over **sprite** in the **for each sprite** block and dragging in its **get sprite** block ❺ and its second socket by snapping in an empty string block ❻. These blocks clear the game board of all Xs and Os when the player hits the button.

Resetting the Label and Global Variables

Now let's add the blocks that will restore **Label1** and the **player** and **play** variables to their original state.

Click **Label1** in the Blocks pane, drag its **setLabel1.Text** block ❶ in under the **for each** loop, and snap an empty string block ❷ into its socket. Then drag a **set global player** block ❸ from the Variables drawer, snap it under the **setLabel1.Text** block, and place another string block ❹ filled with an X into its socket. Finally, drag a **set global play** block ❺ from the Variables drawer, snap it under the **set global player** block, and place a **0** number block ❻ into its socket. These blocks make the **Label** text disappear and set **player** and **play** to their original X and 0 values.

Altogether, the blocks for step 3 direct the app to reset when a player hits Reset Game, just as we planned.

TESTING THE APP

Now we can test the completed app! When you look at your device, you should still see the game board and the reset button and be able to play the game just as you did when you last tested. And now, when you hit the reset button, all the X and O graphics should disappear along with the label text.

Also, behind the scenes, the **player** and **play** variables should reset as well, which you can test by beginning to play the game again. When you touch a square, an X graphic should appear and you should be allowed to touch all nine empty squares before the label displays "Game's over!"

If your reset button doesn't work as planned, debug and try again. Also reset your connection to AI2 Companion if necessary by clicking **Connect ▸ Reset Connection** in the top menu bar and then clicking **Connect ▸ AI Companion** and scanning the QR code. If you placed your blocks correctly, the app should work as expected, and you've successfully created the "Tic Tac Toe" game app!

SUMMARY

In this chapter, you built the "Tic Tac Toe" app, which uses the App Inventor `Canvas` and `ImageSprite` components to create a Tic Tac Toe game for two players. As you built the app, you learned how to create a list of components, access the components' properties using Any component blocks, and iterate through the list with `for` each loops to set up and help control the flow of the game. You worked more extensively with `if then` and logical and comparison operator blocks to control app flow as well.

In the next chapter, you'll learn how to make your code reusable by writing *procedures*—called *methods* and *functions* in some programming environments—using the Procedures blocks. You'll create and call procedures in your code for the "Multiplication Station II" app, which will allow users to choose from one of two difficulty levels of timed multiplication problems.

ON YOUR OWN

Save new versions of "Tic Tac Toe" as you modify and extend it for these exercises. You can find solutions online at *https://nostarch.com/programwithappinventor/*.

1. Extend the game so that it indicates when a player has won the game, meaning a player has placed Xs or Os in three squares in a row, either horizontally, vertically, or diagonally.

2. Extend the app further so that it uses an additional component to indicate which player has won the game.

MULTIPLICATION STATION II: REUSING CODE WITH PROCEDURES

Sometimes you'll want to reuse a section of code to complete a specific task you've completed before. In these situations, you can define or create a *procedure* that includes those lines of code, and then give that procedure a unique name.

This way, whenever you want to execute that task, you can call the procedure by name instead of writing the same lines of code over and over again. This also means that when you have to correct something in that code, you only have to fix the code inside the procedure once.

In this chapter, you'll learn how to create procedures in App Inventor and use them to make the "Multiplication Station II" app. As you'll see, procedures not only help you avoid repeating the same code, they also make long and complicated code more organized and easier to read.

PROCEDURES CAN TAKE PARAMETERS

Let's use a simple example to see how procedures work. Say you've written code that calculates how many days there are until someone's birthday. You can use those lines of code to create a procedure and give it a name like BdayCountDown. Now, whenever you need to count down to someone's birthday, all you have to do is *call* the BdayCountDown procedure. Pretty handy!

Like the built-in methods we've used in some of our apps, a procedure might require *parameters* that you'll have to declare and name when defining the procedure itself. In the birthday example, the BdayCountDown procedure would take a person's birth month and day as parameters to calculate the result. This means that when calling BdayCountDown, you'll need to provide the *arguments* that the procedure needs to complete that task. In this case, the procedure would take numbers (for the date of birth) as its arguments.

Since the arguments passed into the procedure at each call may be different, a procedure might produce a different result every time you call it. This makes sense in the birthday example, too: since most people have different birthdays, the procedure would return different values.

CREATING YOUR OWN PROCEDURE

You've already used several of App Inventor's built-in procedures, both with and without parameters. For instance, in "Hi, World!" in Chapter 1, you used the call to the built-in SpeechRecognizer.GetText method to convert spoken messages to text. And in Chapter 3's "Fruit Loot" app, you moved the ImageSprites around the Canvas using the ImageSprite.Move method call, which required x- and y-coordinates as parameters.

In addition to built-in methods, App Inventor provides blocks that let you create your own procedures, which can return results or not and can have numerous parameters or none at all. Once you create a procedure with its own unique name, App Inventor creates a block just for that procedure that you can use to call it, complete with sockets to plug in any parameter arguments. You'll find the call block in the Procedures blocks drawer.

Let's try creating our own procedures to change the code in some of the apps we've already made. We'll refactor the code in "Fruit Loot" so that we don't repeat the same blocks in the three EdgeReached event handlers that tell the app what to do after each fruit ImageSprite reaches the edge of the Canvas.

The following code shows the EdgeReached event handlers we coded in Chapter 3.

As you can see, we duplicated blocks to create the three separate Edge Reached event handlers for FruitSprite1, FruitSprite2, and FruitSprite3. In each copy, we changed only the few blocks that set the ImageSprite's X property to make sure that the fruit ImageSprites never collide with each other as they fall down the Canvas.

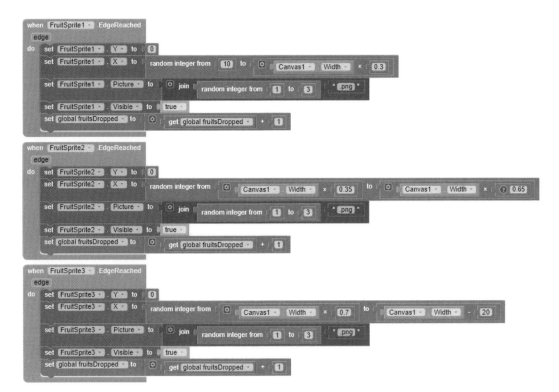

Now compare this with the following code, which shows a backToTop procedure that completes the same tasks, but uses a lot fewer blocks if we call it within the event handlers for all three fruit ImageSprites.

To create the backToTop procedure, we'll need to define the parameters sprite and X, which are the two pieces of information that may change each time the procedure is called—the ImageSprite to adjust and the x-coordinate for the point at the top of the screen where we want to move that ImageSprite.

DEFINING THE PARAMETERS

Log into App Inventor, and open your "Fruit Loot" app by selecting **Projects ▸ My Projects** and choosing the "Fruit Loot" app from your list of projects.

Go to the Blocks Editor and click the **Procedures** blocks drawer in the Blocks pane. Drag the **to procedure do** block to the Viewer, click **procedure**, and rename it by entering **backToTop**.

Then click the blue mutator icon to the left of the words to backToTop, drag two **input:x** blocks to the **inputs** block in the dialog that opens, and rename them **input:sprite** and **input:X**. Your inputs should look like Figure 7-1. Once the mutator dialog closes, you also should see the parameters sprite and X listed to the right of the words to backToTop in the to backToTop block, as shown in the previous code.

Figure 7-1: The blocks that add the parameters
for the backToTop procedure

Now we can program backToTop so that it can have the app execute the exact same commands that we coded in each of the original EdgeReached handlers. But instead of using blocks for a specific ImageSprite, we'll use the generic Any ImageSprite setter blocks. That way, each time we call the procedure, the setter blocks will make changes to the ImageSprite referred to in the procedure's sprite parameter, which will be the ImageSprite that hit the edge.

To add the setter blocks, click the plus sign to the left of the Any Component category listing at the bottom of the Blocks pane and then click the **Any ImageSprite** drawer, drag the **setImageSprite.Enabled** block to the Viewer, and make three copies of it. You should now have four copies of the set ImageSprite.Enabled block on the Viewer. You'll use these to set the Image Sprite's Y, X, Picture, and Visible properties as shown earlier.

Setting the ImageSprite's Y and X Properties

We'll start by setting the given ImageSprite's Y and X properties to make the ImageSprite return to a random spot at the top of the screen before dropping again.

These blocks first take the Y property of the ImageSprite that is given as the argument for the sprite parameter in the backToTop procedure call (the ImageSprite that dropped and hit the edge) and set it to 0. This moves the ImageSprite back up to the top of the screen. Drag the first copy of the **setImageSprite.Enabled** block ❶ inside the **to backToTop** procedure block next to the word do, click the drop-down arrow to the right of the word Enabled, and replace **Enabled** by selecting Y. Then mouse over the procedure's sprite parameter and drag its **get sprite** block ❷ into the **setImageSprite.Y** block's first socket. Drag the number 0 block ❸ from the Y property setter block in the original **FruitSprite1 EdgeReached** handler into the second socket.

Now, we'll add blocks that set the given ImageSprite's X property to the given random X value, which moves the ImageSprite to a random spot at the top of the screen. To do this, drag the second **setImageSprite.Enabled** block ❹ into the procedure block under the **setImageSprite.Y** block, and replace **Enabled** by selecting X in the drop-down. Drag another **get sprite** block ❺ into the **setImageSprite.X** block's first socket, mouse over the procedure's X parameter, and drag its **get X** block ❻ into the second socket.

Setting the ImageSprite's Picture and Visible Properties and Adding to fruitsDropped

Next, we'll add blocks that set the given ImageSprite's new Picture and Visible properties and increment the value of the fruitsDropped variable.

First, we'll place the blocks that set the given ImageSprite's Picture property randomly to *1.png, 2.png,* or *3.png,* each of which is a picture of a different piece of fruit. This ensures that players won't know whether an apple, lemon, or orange will drop next. To do this, drag the third **setImageSprite .Enabled** block ❶ into the procedure block under the **setImageSprite.X** block, and replace **Enabled** by selecting **Picture** in the drop-down. Drag another **get sprite** block ❷ into its first socket, and drag the **join** block ❸ from the Picture setter block in the original EdgeReached handler into the second socket.

Then, we'll add another set of blocks to make the given ImageSprite visible in case it hit the picker earlier and disappeared. Drag the fourth **setImageSprite.Enabled** block ❹ into the procedure block under the **set**

`ImageSprite.Picture` block, and replace **Enabled** by selecting **Visible** in the drop-down menu. Drag another **get sprite** block ❺ into its first socket, and drag the **true** block ❻ from the `Visible` setter block in the original `EdgeReached` handler into the second socket.

Finally, to complete the `backToTop` procedure, we'll add blocks that increment the value of the `fruitsDropped` variable by 1 when the procedure is called. This will enable us to keep track of the total number of pieces of fruit dropped in the game. To do this, drag the **set global fruitsDropped** blocks ❼ from the original `EdgeReached` handler into the procedure block under the `setImageSprite.Visible` block.

CALLING THE PROCEDURE

With this procedure in place, we no longer need to add the five setter blocks to each `EdgeReached` event handler.

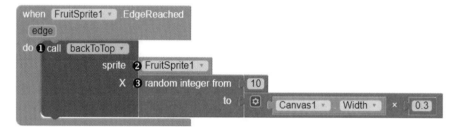

We can handle the `EdgeReached` event for each fruit `ImageSprite` simply by snapping the **call backToTop** block ❶ from the Procedures drawer inside each `ImageSprite`'s `EdgeReached` event handler.

We can then click the **ImageSprite** in the Blocks pane; drag the very last block, its getter block ❷, into the **call backToTop sprite** socket to provide the argument for the sprite parameter; and drag the random integer blocks ❸ from the **X** setter block in its original `EdgeReached` handler into the **call backToTop X** socket to provide the argument for the X parameter. Once you add the call block to the `EdgeReached` handler for each `ImageSprite` and fill the parameter sockets as described, be sure to delete the five original setter blocks from all three `EdgeReached` event handlers.

Now we can use the `backToTop` procedure not just in the "Fruit Loot" game, but in any other similar game. Since we've created this procedure here, we never have to figure out again how to keep moving randomly dropping, non-colliding images back to the top of a screen to drop again, whether they're images of balls, birds, or anything else.

NOTE *You can transfer a procedure, or any other block, from one app project to another by right-clicking on the blocks and selecting Add to Backpack. This action adds a copy of the blocks to the backpack located at the top right of the Viewer in the Blocks Editor window. To take blocks out of the backpack, click the backpack and drag the blocks to the Viewer.*

BUILDING THE "MULTIPLICATION STATION II" APP

Let's define another multiparameter procedure to build on the original "Multiplication Station" app from Chapter 4. When we're done, users should be able to select from two difficulty levels of multiplication problems. We'll also rework the user interface of the app's welcome screen, create new global list variables to present choices, and include an `if then` block to tell the app what to display when a user selects each level of practice problems.

To get started, open your original "Multiplication Station" app by selecting **Projects ▸ My Projects**, as shown in Figure 7-2, and choosing the "Multiplication Station" app from your list of projects.

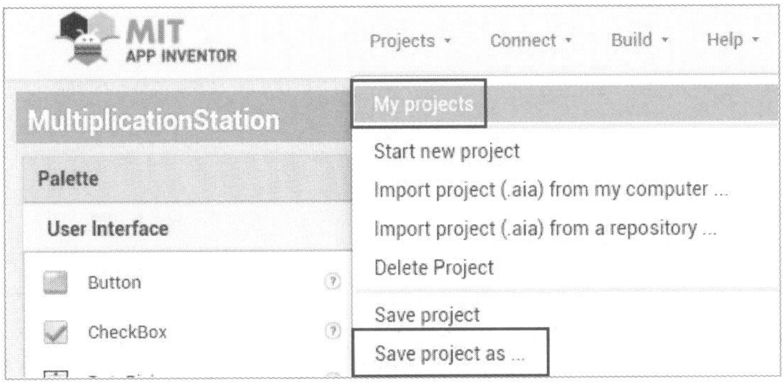

Figure 7-2: The Projects menu where you open and save projects

Once "Multiplication Station" opens, select **Projects ▸ Save project as...** and rename the project by entering `MultiplicationStationII` without any spaces, and then click **OK**.

DECOMPOSING "MULTIPLICATION STATION II"

We want to change the original "Multiplication Station" app so that when users open the app, they can choose level 1 (easier) or level 2 (more challenging) problems to practice and will have more time to solve level 2 problems.

We can decompose the new action for "Multiplication Station II" into three steps:

1. When the user opens the app, play a welcome message. Display the `ListPicker` for the user to click to choose a practice level.
2. After the user clicks the `ListPicker` to choose a practice level, play another message stating the number of seconds available to answer each problem. Open and pass the selected problem level to the practice screen.
3. When the `Clock`'s timer fires, display a random multiplication problem at the correct time interval for the selected problem level.

You'll need the following new components:

- **ListPicker** for the user to click to select problem level and open the practice screen (this replaces the original start button on the welcome screen)
- Procedure to display problems and set the Clock timer interval according to the selected problem level
- Variable (2) to store problem level and timer interval

LAYING OUT "MULTIPLICATION STATION II" IN THE DESIGNER

Right now, the welcome screen gives users only one option—to click the start button and start practicing. To allow users to first choose the level of multiplication problems, all we have to do is remove Button1 from the welcome screen and replace it with a ListPicker. Once we adjust the ListPicker's properties to make it look similar to Button1, the new welcome screen should look something like Figure 7-3.

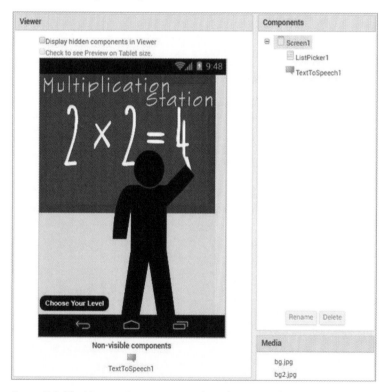

Figure 7-3: The Viewer, Component, and Media panes showing the Screen1 layout for "Multiplication Station II"

To make these changes, first go to the Designer for Screen1. Click Button1 in the Components pane, and click **Delete** and then **Delete** again in the dialog that opens.

Now drag a `ListPicker` from the User Interface drawer and place it on the Viewer where you just deleted `Button1`. Adjust the `ListPicker`'s properties so that it looks like `Button1`. In the Properties pane, change the background color to black by clicking **Default** under `BackgroundColor` and then clicking **Black** when the color list dialog opens. Make its text bold by clicking the checkbox under `FontBold`, and change its shape by clicking the drop-down arrow under **Shape** and selecting **rounded**.

Next, enter `Choose Your Level` in the text box under **Text** so users know to click the `ListPicker` to choose their problem level. Then, center the text by clicking the drop-down arrow under `TextAlignment` and selecting **center: 1**, and make the text white by clicking **Default** under `TextColor` and then clicking **White** when the color list dialog opens. `Screen1` should now look like Figure 7-3 in the Viewer.

PROGRAMMING "MULTIPLICATION STATION II"

Now let's program the new features for "Multiplication Station II" following the steps outlined earlier. To show the correct problems for the selected problem level, we'll create two new list variables and a procedure with parameters. We'll add an `if then` block to the procedure so that the app will know whether to display problems for level 1 or level 2 at each `Clock` timer interval.

We'll also program two new event handlers so that the app will know what to do before and after the user selects a level with the `ListPicker`. To begin programming, click the **Blocks** button to switch to the Blocks Editor and make sure you're on `Screen1`.

STEP 1: SETTING PROBLEM LEVEL CHOICES AND PLAYING THE NEW WELCOME MESSAGE

As soon as the screen opens, we want to display the `ListPicker` with the problem level options for the user to select. We'll prompt the user to select a practice problem level by playing the app's welcome message.

Creating the Global level and seconds List Variables

We'll use two global list variables in this step: `level`, which stores the two problem level choices for `ListPicker1`, and `seconds`, which holds the number of seconds available to solve problems in each level. Create the two lists with the blocks shown here.

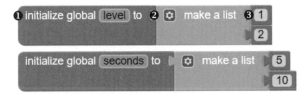

To create each variable, click the **Variables** block drawer and drag an `initialize global name` block ❶ to the Viewer, click **name**, and replace it with the variable name (in this case, **level** and **seconds**). Then drag a make a list block ❷ from the Lists drawer and snap it to the right side of the **initialize global** block. Finally, drag two **0** number blocks ❸ from the Math drawer and snap them into the sockets of the make a list block. Then, replace the **0** in each number block with **1** and **2** in **level** for the two possible problem levels and **5** and **10** in **seconds** for the number of seconds available for each level.

The items we've placed in the same index position in the two variables correspond to each other. This means that we can see how many seconds are available to solve problems for each level by looking at the items that are placed in the same position in the **level** and **seconds** variables. Here you can see that there are 5 seconds available for level 1 problems, and 10 seconds available for level 2.

Programming the Welcome Screen Event Handlers

Now we can start coding the app's response to the two events in this step: the ListPicker BeforePicking and the ScreenInitialize events. This is where we want to set the problem level choices for the ListPicker and play the welcome message.

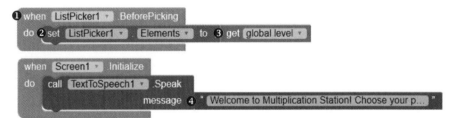

Having created **level**, we have the data we need to set ListPicker1's choices. Before the user selects from ListPicker1, we need to set those choices to the items in the **level** list so users can choose between level 1 and level 2. We'll program this action with the ListPicker BeforePicking event handler.

In the Blocks pane, click **ListPicker1** and drag the **whenListPicker1.Before Picking** event handler block ❶ to the Viewer. Then, click **ListPicker1** again and drag its **setListPicker.Elementsto** block ❷ into the **whenListPicker1.Before Picking** block next to the word do. Next, in the Blocks pane, click the **Variables** blocks drawer, drag the **get global level** block ❸ in, and snap it to the right of the **setListPicker1.Elementsto** block. These three blocks create the Before Picking event handler, which sets the problem level choices that the user will see when clicking ListPicker1.

Finally, for the whenScreen1.Initialize event handler, which we already dragged to the Viewer for the original "Multiplication Station" app, we can leave all blocks the same, except for the string block that holds the argument for the message parameter of the callTextToSpeech1.Speak block. In that string block ❹, change the text to read **Welcome to Multiplication Station! Choose your problem level to begin practicing.** so users will hear that message when the app opens.

Live-test these event handlers with a device. When your "Multiplication Station II" app opens on your device, you should hear the welcome message.

You also should see the Choose Your Level button on the screen, and when you click it, you should see the two choices, 1 and 2. Right now, if you try to select either choice, nothing should happen. We'll program that AfterPicking action next.

If the app's not working as described, debug and make sure you've created and placed your blocks correctly. When everything's working, leave the app open on your device to keep live-testing.

STEP 2: STATING THE NUMBER OF SECONDS FOR EACH PROBLEM AND OPENING THE PRACTICE SCREEN

Let's use the ListPicker AfterPicking event to tell users how many seconds they have to answer each problem, depending on the level they selected. Then we'll program the app to open the practice screen and transfer the value of the selected problem level to that screen. Here are the blocks that direct this action.

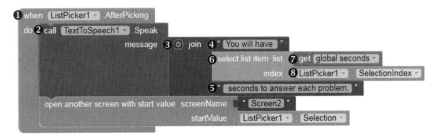

Let's first program the app to speak after the user chooses a problem level.

Telling the User the Time Limit

In the Blocks pane, click **ListPicker1** and drag the **whenListPicker1.AfterPicking** event handler block ❶ to the Viewer. Then, click **TextToSpeech1** and drag its **callTexttoSpeech1.Speak** block ❷ into the **whenListPicker1.AfterPicking** block next to the word do.

Next, to provide the argument for the callTexttoSpeech1.Speak block's message parameter so the app will know what to say, drag a **join** block ❸ from the Text blocks drawer to the Viewer. Then add another string input to the **join** block and snap it to the right of the **callTexttoSpeech1.Speak** block next to the word message. This lets us join three strings together to form the message for TextToSpeech1. If the user selected problem level 1, we fill the join block's inputs to set the message to **You will have 5 seconds to answer each problem.**, and if the user selected problem level 2, we set the message to **You will have 10 seconds to answer each problem.**

To fill the **join** block's top and bottom inputs, drag in two empty string blocks from the Text blocks drawer, click their text areas, and enter **You will have** (with a space after the word have) in the top string block ❹ and **seconds to answer each problem** (with a space before the word seconds) in the bottom ❺.

Then, for the join block's middle input, we need to add the number of seconds the user will have to answer each problem. We'll get that value by directing the app to select the item from the seconds list that holds the same index position as the user's ListPicker selection in the level list.

To program this, drag in a **select list item** block ❻ from the List blocks drawer and snap it into the **join** block's middle input. Then, identify the list from which we want the app to select the item by clicking the **Variables** blocks drawer and dragging the **get global seconds** block ❼ into the **select list item** block's **list** socket. Finally, identify the index position of the item selected from the seconds list by clicking **ListPicker1** and dragging the **List Picker1.SelectionIndex** block ❽ into the **select list item** block's **index** socket.

Moving to the Practice Screen

Once the user selects a problem level, we'll open Screen2, the practice screen, and pass the value the user selected in the ListPicker to that screen.

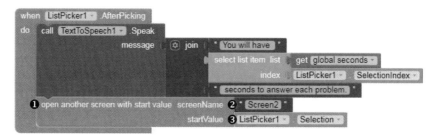

In the Blocks pane, click the **Control** blocks drawer, drag in the **open another screen with start value** block ❶, and snap it inside the **whenListPicker1 .AfterPicking** block under the callTexttoSpeech1.Speak block.

This open another screen with start value block requires us to provide a screenName, the name of the screen we want to open (Screen2), and the start Value to pass to Screen2 (the problem level the user selected in ListPicker1). To provide these values, in the Blocks pane, click the **Text** blocks drawer, drag in another empty string block ❷, and snap it into the screenName socket. Enter **Screen2** inside the empty string block to provide the name of the screen we want to open.

Then, to identify the start value we want to pass to Screen2, click **List Picker1**, drag in its **ListPicker1.Selection** block ❸, and snap it into the start Value socket.

Test to see how these blocks work. Once you open the app, you should hear the welcome message and see the Choose Your Level button. Once you click the button, you should see the numbers 1 and 2 as your choices. When you click the number 1, you should hear the app say you have 5 seconds to answer each problem, and when you click 2, it should say you have 10 seconds. Then the app should take you to the practice screen.

Let's wait to test whether the app properly transfers the start value to the practice screen until after we program the next step. Make sure everything else you've programmed in the app is working as described, and leave it open on your phone for more live-testing.

Just like the original "Multiplication Station" app, this app displays multiplication problems by joining one random integer (the value of global variable a), the multiplication operator (×), and another random integer (the value of global variable b). The app also gives users a set time interval to answer each problem. In this step, we'll create the setLevel procedure that sets the values of a and b and the Clock timer interval value in each procedure call, depending on the problem level the user selected.

We'll make the setLevel procedure require arguments for five parameters: a1, a2, b1, b2, and interval. Parameters a1 and a2 set the random integer range for global variable a, the first random number in each multiplication problem, and parameters b1 and b2 set the random integer range for global variable b, the second random number. The interval parameter holds the time interval in milliseconds. Once setLevel receives these arguments in a procedure call, it sets global variable a to a random number ranging from a1 to a2, sets global variable b to a random number ranging from b1 to b2, and sets the Clock timer interval to interval.

Here's what the setLevel procedure looks like.

We'll call setLevel inside an if then block to place the correct numbers in the multiplication problems and give the user the right amount of practice time based on the selected problem level.

Creating the setLevel Procedure

To make setLevel, click the **Procedures** block drawer in the Screen2 Blocks pane, drag a **to procedure do** block ❶ to the Viewer, and click **procedure** to rename it by entering **setLevel**. This procedure block, like the global variables, will stand alone outside of all event handlers, allowing us to call the procedure within each event handler if we need to.

Next, list the required procedure parameters by clicking the blue mutator icon to the left of the words to setLevel and dragging five **input:x** blocks to the **inputs** block in the dialog that opens. Rename them **input:a1**, **input:a2**, **input:b1**, **input:b2**, and **input:interval**.

Now let's program the action setLevel will perform. We'll start with the blocks that set global variable a to a random number ranging from the argument provided for a1 to the argument provided for a2. Click the **Variables** blocks drawer and drag a **set global a** block ❷ into the **to setLevel** block next to the word do. Then drag a **random integer** block ❸ from the Math drawer and snap it to the right of the **set global a** block. Fill the **random integer** block's first socket by mousing over the procedure's **a1** parameter and dragging in a **get**

a1 block ❹, and fill its second socket by mousing over the **a2** parameter and dragging in a **get a2** block ❺.

To set global variable **b** to a random number ranging from **b1** to **b2**, simply repeat the process above for global variable **b** by copying the blocks at ❷ through ❺, and snapping the copy under the **set global a** block. Use the drop-down arrows to change **global a** to **global b**, **a1** to **b1**, and **a2** to **b2**.

Finally, to complete the setLevel procedure, we'll set the Clock's Timer Interval to the argument provided for interval in the procedure call. Click **Clock1** in the Blocks pane and drag its **setClock1.TimerIntervalto** block ❻ into the **to setLevel** block under the **set global b** block. Then mouse over the procedure's **interval** parameter and snap its **get interval** block ❼ to the right of the **setClock1.TimerIntervalto** block.

Calling setLevel

We can now call setLevel in the ClockTimer event handler so that the app displays the correct problems for each level at the right time. The following code shows how we'll call setLevel in a new if then else block within the whenClock1.Timer block, providing one set of arguments for the required parameters if the start value passed in from Screen1 equals 1 and another set of arguments if it doesn't.

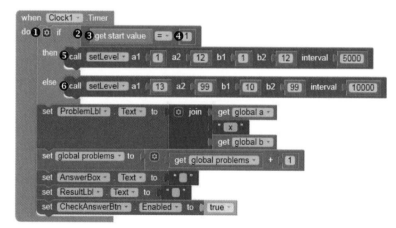

Here, we change the original "Multiplication Station" app's whenClock1 .Timer event handler block by replacing the global variable a and global variable b setter blocks at the top with an if then else block. Delete the two setters, drag an **if then** block ❶ to the Viewer from the Control blocks drawer, click the blue mutator icon to add an **else** input, and snap the **if then else** block inside the **whenClock1.Timer** block next to the word do.

Now we'll add blocks to test whether the start value passed in from Screen1 equals 1, which means that the user selected problem level 1 in the ListPicker on Screen1. Drag an = comparison block ❷ from the Math drawer and snap it into the **if then else** block's if socket. Fill the = comparison block's left operand socket with a **get start value** block ❸ from the Control drawer and its right operand socket with a **1** number block ❹.

The next blocks tell the app how to respond to the answer to that test condition, which is whether the start value equals 1. If the answer is yes, we want the app to set the correct problems and time interval for level 1 problems. If the answer is no, we want the app to set the correct problems and time interval for level 2 problems. In each instance, we'll direct the action with a call to the setLevel procedure.

To add the blocks for these procedure calls, click the **Procedures** drawer, drag one **call setLevel** block ❺, and snap it into the **if then else** block's **then** socket. Then drag another **call setLevel** block ❻ and snap it into the **if then else** block's **else** socket. Next, fill the parameter arguments for each call block with the values shown in Table 7-1.

Table 7-1: Setting Parameters for Calls to setLevel

if then else **block socket**	a1	a2	b1	b2	interval
then	1	12	1	12	5000
else	13	99	10	99	10000

Each time the Clock timer fires, if the start value passed from the welcome screen equals 1 (then, user chose problem level 1), the app calls the setLevel procedure to set global variable a to a random number ranging from 1 to 12, set global variable b to a random number ranging from 1 to 12, and set the Clock timer interval to 5 seconds.

If the start value passed from the welcome screen doesn't equal 1 (else, user chose level 2), the app calls setLevel to set global variable a to a random number ranging from 13 to 99, set global variable b to a random number ranging from 10 to 99, and set the Clock timer interval to 10 seconds.

Now let's switch to Screen1 and test the completed app! When the welcome screen opens in your device, you should see the ListPicker and hear the welcome message. When you click Choose Your Level and make a selection, the app should tell you how many seconds you have to answer each problem. The practice screen should also open, and, if you chose level 1, problems should appear every 5 seconds, or, if you chose level 2, problems should appear every 10 seconds. Also, the problems should include numbers from the range expected for the selected problem level.

If any part of the app isn't working as planned, debug and try again. If you placed your blocks correctly, the app should work as expected, and you've successfully created "Multiplication Station II"!

SUMMARY

In this chapter, you built the "Multiplication Station II" app, which builds on the original "Multiplication Station" app you created in Chapter 4 to let users choose the difficulty level of the problems and pass that choice from one screen to the next. Instead of having to use the same code blocks over and over again, you learned how to use a procedure to reuse the code that sets the problems and time limit. You also worked more with lists, if then, and relational operator blocks to control app flow.

In the next chapter, you'll learn how to create apps that let users draw on the screen and drag images around. You'll create the "Virtual Shades" app, where a user can take a selfie, "try on" a variety of sunglasses by dragging them into place over the picture, and draw and type on the picture with different-colored "ink."

ON YOUR OWN

When you're ready for a challenge, save new versions of your previous apps, and try these exercises. You can find solutions online at *https://nostarch.com/programwithappinventor/*.

1. In "Multiplication Station II," try adding a pause button to the practice screen so users can temporarily stop practicing when they need to. Write and call a procedure as part of this extension.

2. Write and call a procedure that changes the code you wrote in Chapter 4 to allow users to choose to practice multiplication or division problems.

VIRTUAL SHADES: DRAWING AND DRAGGING IMAGES

For the "Fruit Loot" and "Tic Tac Toe" games in Chapters 3 and 6, you animated the Canvas and ImageSprite components from App Inventor's Drawing and Animation drawer by programming EdgeReached, TouchDown, and other event handlers.

In this chapter, we'll program Canvas and ImageSprite moving, drawing, and dragging methods and event handlers to create an app that lets users draw, type, and drag images across a smartphone or tablet screen. Developers use similar functions to build apps that allow users to move elements, take handwritten notes, and doodle on the screen.

BUILDING THE "VIRTUAL SHADES" APP

The "Virtual Shades" app lets users take pictures of themselves with their cameras and then drag images of sunglasses over their selfies to try them on virtually. As you build the app, you'll learn to use the Camera to create the

Canvas background, the Slider to adjust other components, and the Notifier to collect user input. You'll also practice using many programming concepts you've already learned.

To get started, log in to App Inventor and create a new project named VirtualShades.

DECOMPOSING "VIRTUAL SHADES"

For this app, users should be able to take a picture that becomes the Canvas background, drag various pairs of sunglasses over the picture, and change the size of the sunglasses to get the right fit. They also should be able to doodle and add text to the photo; save it as a graphic that they can view, email, or text later; and erase the sunglasses as well as all drawing and text at any time.

We can break this action into eight steps:

1. When the user clicks the picture button, open the device's camera. Set the picture the user takes as the Canvas background.
2. When the user clicks the ListPicker, show the list of available sunglasses.
3. After the user selects a pair of sunglasses, display them on the Canvas.
4. When the user drags the sunglasses around the Canvas, move them where the user drags. When the user moves the Slider's thumb, adjust the size of the sunglasses.
5. When the user clicks the draw button, show a Notifier alert letting the user know where to draw. When the user drags a finger around the Canvas, draw on the Canvas where the user drags.
6. When the user clicks the type button, show a Notifier with a text box where the user can enter text. After the user submits the text, display it at the bottom of the Canvas.
7. After the user clicks the save button, store a picture of the Canvas on the device, and show a Notifier alert letting the user know the storage location.
8. When the user clicks the trash button, erase all sunglasses, marks, and typing from the Canvas.

You'll need the following components:

- **Button** (5) for the user to click to open the camera, draw, type, save, and erase the Canvas
- **Camera** to open the device's camera
- **Canvas** to enable dragging and drawing
- **HorizontalArrangement** (2) to hold the Label, Slider, and Buttons
- **ImageSprite** (4) for the user to drag around the Canvas
- **Label** to provide Slider directions
- **ListPicker** for the user to click to choose sunglasses ImageSprites
- **Notifier** to show alerts and capture user text input

- Procedure to clear images from the Canvas and reset the Slider
- Slider for the user to adjust the size of ImageSprites
- Variable (2) to store the ImageSprites and their descriptions
- VerticalArrangement to hold all visible components except the picture Button

LAYING OUT "VIRTUAL SHADES" IN THE DESIGNER

Now let's start laying out the app in the Designer by adjusting the app screen.

Adjusting the Screen

First, change the screen's horizontal alignment to center all contents. Then, to give users the most space to "try on" sunglasses, make the screen stay in portrait mode even when the user rotates the device. To do this, select **Portrait** under ScreenOrientation. Also, hide the status bar and screen title so they won't take up space by unclicking the checkboxes under both ShowStatusBar and TitleVisible.

Adding the Canvas and ImageSprites

Let's place most of the visible components on the screen in a Vertical Arrangement so that we can set their visibility as a group. When the app opens, only the Button to take a picture should show; the other components will go in the VerticalArrangement so we can hide them until later.

Drag a **VerticalArrangement** from the Layout drawer onto the Viewer. In the Properties pane, click the drop-down arrow under AlignHorizontal and select **Center: 3**, and set its height and width to **Fill parent** to make it cover the entire screen.

Now add components to the VerticalArrangement in the order in which we want them to appear. Start by dragging a Canvas from the Drawing and Animation drawer onto the Viewer inside of VerticalArrangement1. In the Properties pane, remove the Canvas's background color by clicking the box under BackgroundColor and selecting **None**. Change its text size to 25 in the text box under FontSize. Make its height and width **Fill parent**, and increase the width of the line that can be drawn on it to 4 in the text box under LineWidth. Remove its paint color by clicking the box under PaintColor and selecting **None**. Lastly, left-align its text by clicking the drop-down arrow under Text Alignment and selecting **left: 0**.

Now, drag four ImageSprites from the Drawing and Animation drawer onto the Canvas, click them in the Components pane, and rename them Sunglasses1, Sunglasses2, Sunglasses3, and Sunglasses4. Upload the images *tanSunglasses.png*, *purpleSunglasses.png*, *redSunglasses.png*, and *blackSunglasses .png*, all of which come with the resources for this book, to the ImageSprites by clicking **Picture** in the Properties pane for each. Also, make each Image Sprite invisible by deselecting the box under **Visible**, and change both the X and Y property for each to 0 so that they appear at the top left of the Canvas when they show up in the app.

Adding User Buttons and Controls

Drag a `ListPicker` from the User Interface drawer into `VerticalArrangement1` under the Canvas and make its background color red, its font bold, its height 35 pixels, and its width **Fill parent**. Also, change the default text to **Choose Glasses** and make the text white.

Now we'll add a `Label` and `Slider`, which we want to position side by side to save space. That means we need to place them within a `HorizontalArrangement`. Drag a `HorizontalArrangement` from the Layout drawer into `VerticalArrangement1` under the `ListPicker` and rename it **SliderArrangement**. Then, select **Center: 3** under **AlignHorizontal** and make its width **Fill parent**.

Drag a `Label` and a `Slider` from the User Interface drawer into `Slider Arrangement`, with the `Label` on the left. Make the `Label`'s font bold, change its default text to **Adjust Sunglasses Width:**, center-align its text, and make the text red.

Make `Slider1`'s **ColorLeft** red to match the color scheme and **Width** 40 percent so it will fit next to the `Label`. Also, so the user can change the width of the sunglasses from 150 to 250 pixels, make the `Slider`'s **MaxValue 250** and **MinValue 150**. Then set the **ThumbPosition** to **200** so the `Slider`'s thumb will sit in the middle of the slider when it appears. Next, disable the `Slider`'s thumb by unclicking the checkbox under **ThumbEnabled**, which will keep the `Slider` from moving until we program it to work after a user drags sunglasses onto the Canvas.

Now let's add the Buttons. We'll place the first four Buttons side by side in a second `HorizontalArrangement`, also to save space. Place this `Horizontal Arrangement` under **SliderArrangement** in `VerticalArrangement1` and name it **ButtonArrangement**, center its contents by selecting **Center: 3** under **Align Horizontal**, and make its width **Fill parent**.

Drag four buttons from the User Interface drawer into **ButtonArrangement** and rename them, from left to right, **DrawBtn**, **TypeBtn**, **SaveBtn**, and **TrashBtn**. Then, in the Properties pane for each, under **Image**, upload *DrawBtn.png*, *TypeBtn.png*, *SaveBtn.png*, and *TrashBtn.png*, respectively, all of which come with the resources for this book. Set the height of each `Button` to 50 pixels and remove each `Button`'s default text from the text box under **Text**. Note that the four buttons may not be fully visible in the Viewer, but you should see them when you test on a device.

Now that we've placed all the components we need inside of `Vertical Arrangement1`, we need to hide it until after the user takes a picture. To make it invisible, in its Properties pane, unclick the checkbox under **Visible**.

Finally, add the last `Button`, which is the only visible component we'll place outside of `VerticalArrangement1`. Users click this `Button` to open the camera when they first enter the app. Drag the new **Button** from the User Interface drawer onto the Viewer, and rename it **TakePicBtn** in the Components pane. Make its height and width **Fill parent** to cover the entire screen. Next, add a background image to the `Button` by clicking the text box under **Image** and uploading *TakePictureBtn.png*, which comes with the resources for this book. Also remove the `Button`'s default text.

Preparing the Non-Visible Components

Now let's add and adjust the non-visible components. Drag a **Camera** component from the Media drawer and a **Notifier** component from the User Interface drawer. Change the **Notifier**'s **BackgroundColor** to red.

At this point, your screen should look like Figure 8-1.

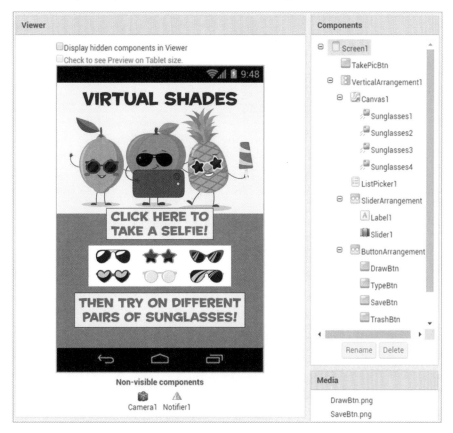

Figure 8-1: The Viewer, Components, and partial Media panes after you've laid out "Virtual Shades"

Now that you've laid out all components, move to the Blocks Editor to program the app.

PROGRAMMING "VIRTUAL SHADES" IN THE BLOCKS EDITOR

For "Virtual Shades," we'll create two list variables and a procedure, and we'll program 15 event handlers with three conditionals and one for each loop.

STEP 1: SETTING THE SELFIE AS THE CANVAS BACKGROUND

In this step, we want the user to take a photo that will become the Canvas background for the user to drag and draw on.

Taking the Photo

We start by having the app open the camera when a user clicks TakePicBtn. Here is the event handler for this.

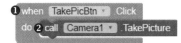

In the Blocks pane, click **TakePicBtn** and drag the **whenTakePicBtn.Click** event handler block ❶ to the Viewer. Then, drag **Camera1**'s **callCamera1.Take Picture** method call block ❷ next to the word do. When the user clicks TakePicBtn, these blocks open the camera so the user can take the photo.

Setting the Canvas Background and Showing App Controls

After the user takes the selfie, we need the picture to become the Canvas background, TakePicBtn to disappear, and the VerticalArrangement that holds the Canvas and other user interface components to appear. The Camera's AfterPicture event handler programs this action.

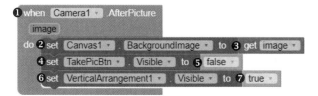

Drag **Camera1**'s **whenCamera1.AfterPicture** event handler block ❶ to the Viewer. This event handler reports the value for the image event parameter, which is the picture taken by the Camera. We'll set the Canvas background to that value.

Drag **Canvas1**'s **setCanvas1.BackgroundImageto** block ❷ into the **whenCamera1 .AfterPicture** block. Then mouse over the **image** event parameter and snap its **get image** block ❸ to the right of the **setCanvas1.BackgroundImageto** block. These blocks set the user's picture as the Canvas background.

Now, to make TakePicBtn disappear, drag its **setTakePicBtn.Visibleto** block ❹ into the event handler block and snap in a **false** block ❺.

Next, we'll make the Canvas and components in the VerticalArrangement visible so the user can drag and "try on" the sunglasses and draw and type on the Canvas. Drag **VerticalArrangement1**'s **setVerticalArrangement1.Visibleto** block ❻ into the event handler and attach a **true** block to it ❼.

Put together, the blocks in step 1 open the camera when the user clicks the picture button and, after the user takes a picture, set the Canvas background to that picture, hide the button by setting its Visible property to false, and show the VerticalArrangement containing the Canvas and app controls by setting its Visible property to true.

To see how these blocks work, live-test with a device. You should see the full-screen TakePicBtn when you open the app. When you click it, your camera should open, and the picture you take should become the background of the top portion of your app screen, where you placed the Canvas. If the picture is rotated the wrong way in the app, close and reopen the app, rotate your camera, and take another photo until it displays the way you want.

Below the picture, you should see the ListPicker with the app's Slider and the draw, type, save, and trash buttons beneath it, but none of these controls should work yet. Leave the app open on your device to keep live-testing.

STEP 2: ADDING SUNGLASSES TO CHOOSE FROM

Now let's program step 2 of the app, the ListPicker's BeforePicking event handler, which will set the options for the ListPicker that the user clicks to select each pair of sunglasses.

To provide these options, we'll create a global variable called glasses Descriptions to hold descriptions of the four pairs of sunglasses. We'll also create a related global list variable, glassesSprites, which will hold the Image Sprites that display the sunglasses. Creating these as global variables allows us to use them in every event handler and procedure in the app.

Creating the glassesDescriptions and glassesSprites List Variables

The following blocks create glassesDescriptions, the variable that holds the options we'll add to the ListPicker to describe each pair of sunglasses, and glassesSprites, the variable that holds the ImageSprites (Sunglasses1 through Sunglasses4) that display the corresponding pictures of the sunglasses.

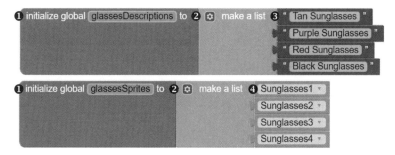

To create the variables, drag two **initialize global name** blocks ❶ from the Variables block drawer to the Viewer and name one variable **glasses Descriptions** and the other **glassesSprites**. Then attach a **make a list** block ❷ from the Lists drawer to each **initialize global** block.

Add two more sockets to each `make a list` block's two default inputs by clicking the blue mutator icon to the left of the words `make a list` and dragging the additional `item` blocks to the `make a list` block in the dialog that opens. Finally, for glassesDescriptions, drag in four empty string blocks ❸, snap them into the sockets of the `make a list` block, and enter the text shown. For glassesSprites, click Sunglasses1 through Sunglasses4 in the Blocks pane and drag each ImageSprite's getter block ❹ into its socket in the `make a list` block. These blocks create the two variables so that each item in glassesSprites corresponds with the item at the same index position in glassesDescriptions.

Setting the ListPicker Options

Next, we'll code the ListPicker's BeforePicker event handler, which sets the ListPicker's elements to the items in glassesDescriptions.

To add the blocks, drag ListPicker1's **whenListPicker1.BeforePicking** event handler block ❶ to the Viewer. Then drag its **setListPicker1.Elementsto** block ❷ inside the event handler block, and snap the **get global glasses Descriptions** block ❸ to the right of the **setListPicker1.Elementsto** block.

Live-test the app to see how these blocks work. When you click the ListPicker, you should see the selection options, which are descriptions of each pair of sunglasses. Nothing should happen yet when you select a pair, except that the ListPicker should close. Leave the app open on your device to keep live-testing.

STEP 3: SELECTING A PAIR OF SUNGLASSES TO TRY

Let's now program the ListPicker's AfterPicking event handler, which tells the app what to do after the user picks a pair of sunglasses. At that point, if there's already a pair of sunglasses on the Canvas, the app should remove it and reset the Slider that adjusts the width of the sunglasses, in case the user previously adjusted it. If, on the other hand, it's the user's first pick from the ListPicker, we want the app to enable the Slider. Then, we want the app to show the pair of sunglasses the user selected.

Creating the clearSprite Procedure

To clear an existing ImageSprite and reset the Slider, we'll use a procedure called clearSprite, which we'll call in the ListPicker's AfterPicking event handler and later in the handler we'll create to clear the Canvas. The clear Sprite procedure uses a conditional block to test whether there's an image on the Canvas. If there is, it removes the image and moves the Slider thumb back to its starting position.

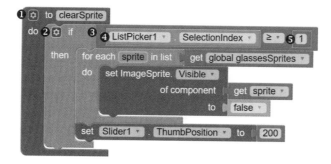

Click the **Procedures** blocks drawer in the Blocks pane, drag the **to procedure do** block ❶ to the Viewer, and name the procedure `clearSprite`. Next, we'll test whether there's an ImageSprite on the Canvas by determining whether the current ListPicker selection index is 1 or greater. Since the four sunglasses ImageSprites have the index positions 1 through 4, we'll know by checking the index whether one of them is showing on the Canvas. If it is, then the app needs to clear it.

Snap an **if then** block ❷ inside the `clearSprite` procedure block and a ≥ comparison block ❸ into the **if then** block's **if** socket. Fill the comparison block's left operand socket with the **ListPicker.SelectionIndex** block ❹ and its right socket with a **1** number block ❺. So far, the **if then** block sets up the test condition by saying, "If the ListPicker1 selection index is greater than or equal to 1" and the sentence stops there. The next blocks complete the sentence by telling the app what to do if the condition is met.

Clearing All ImageSprites and Resetting the Slider

If the ListPicker1 selection index is greater than or equal to 1, the following blocks loop through and hide all ImageSprites and reset the Slider.

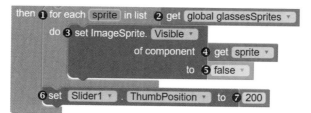

To place the blocks, drag a **for each item** block ❶ inside the **if then** block's **then** socket, and change **item** to **sprite**. Then snap a **get global glasses Sprites** block ❷ to the right of the **for each** block next to the word list. Next, click the plus sign to the left of **Any component** near the bottom of the Blocks pane, click the **Any ImageSprite** drawer, and drag a **setImageSprite .Visible** block ❸ inside the **for each** block. Fill the setter block's top socket by mousing over the **sprite** item and dragging in a **get sprite** block ❹, and fill its bottom socket with a **false** block ❺. Now, when the procedure is called, all ImageSprites will become invisible.

Next we'll add the blocks that reset the Slider. We need to do this because, later in the code, we'll program an ImageSprite's width to change when a user changes the Slider's ThumbPosition. Just in case the user has changed the ThumbPosition for one of the ImageSprites when it was showing on the Canvas, the next blocks change the Slider's ThumbPosition and the ImageSprite's width back to the 200 we set in the Designer.

Drag Slider1's setSlider1.ThumbPosition block ❻ inside the if then block under the for each block, and snap a 200 number block ❼ to its right. Now, when the procedure is called, these blocks reset the Slider.

Calling clearSprite and Enabling the Slider

Now that we've created the procedure, let's build the AfterPicker event handler.

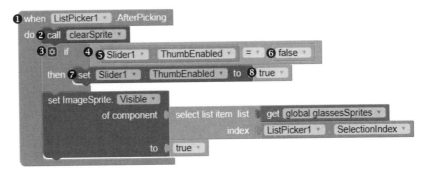

The AfterPicker event handler will first call the clearSprite procedure we just created to clear any ImageSprites from the screen and reset the Slider. It will then enable the Slider, if necessary, and show the selected pair of sunglasses.

Drag ListPicker1's whenListPicker1.AfterPicking block ❶ to the Viewer. Then, drag the call clearSprite block ❷ into the event handler block. These blocks call the clearSprite procedure after the user selects a pair of sunglasses in the ListPicker.

Next, we'll add the conditional block that lets the app know when to turn on the Slider, which we turned off in the Designer by unclicking the ThumbEnabled property. We disabled the thumb so that a user can't move it before selecting a pair of sunglasses, which would cause an error.

It's possible, though, that the user has already enabled the Slider by previously selecting a pair of sunglasses, so we must check for that. In the Blocks pane, drag an if then block ❸ into the event handler under the call clearSprite block. Then, fill the if socket with an = comparison operator block ❹, and fill the operator block's left operand socket by dragging in Slider1's Slider1.ThumbEnabled block ❺. Fill its right operand socket with a false block ❻.

Next, fill the if then block's then socket with a setSlider1.ThumbEnabled block ❼ and drag a true block ❽ to its right. Now, after the user selects a pair of sunglasses, this if then block will enable the Slider's thumb *only if* it's currently not enabled. Although the app will check this condition each time the user selects a pair of sunglasses, it will enable the Slider's thumb only once, after the user's first selection.

Showing the Selected ImageSprite

Finally, the AfterPicker event handler will display the selected pair of sunglasses on the Canvas. When added to the event handler, the following blocks will make this happen.

These blocks set the Visible property of the user's selected ImageSprite to true after they identify the correct ImageSprite by matching its index position in the glassesSprites list with the index of the selected ListPicker1 option.

To place these blocks, click the **Any ImageSprite** drawer and drag a set ImageSprite.Visible block ❶ inside the event handler block under the if then block. Fill the setImageSprite.Visible block's top socket with a select list item block ❷ from the Lists blocks drawer. Next, fill the select list item block's list socket with a get global glassesSprites block ❸ and its index socket with ListPicker1's ListPicker1.SelectionIndex block ❹. Now, fill the setImageSprite.Visible block's bottom socket with a true block ❺.

That's it for step 3. After the user selects a pair of sunglasses, the blocks for this step clear any existing ImageSprite from the Canvas and reset the Slider, enable the Slider if it's the user's first time picking from the ListPicker, and show the selected sunglasses on the Canvas.

Let's live-test the blocks for this step. When you click the ListPicker and make a selection, the sunglasses you selected should show on the Canvas after the ListPicker closes. You won't be able to move the glasses around on the screen yet—not until we program step 4. Leave the app open on the screen to keep live-testing.

STEP 4: TRYING ON SUNGLASSES

In this step, we'll program two event handlers: the ImageSprite Dragged event handler that moves the sunglasses around the Canvas when the user drags them and the Slider PositionChanged event handler that makes the sunglasses bigger or smaller when the user adjusts the Slider.

Setting the Sunglasses1 X- and Y-Coordinates After Dragging

We'll have to program a Dragged event handler that moves the ImageSprites when the user drags the sunglasses around the screen. The following blocks move Sunglasses1.

Drag Sunglasses1's whenSunglasses1.Dragged block ❶ to the Viewer. Then drag its callSunglasses1.MoveTo block ❷ into the event handler block. These blocks call the MoveTo method to move Sunglasses1 to a given point when a user drags Sunglasses1.

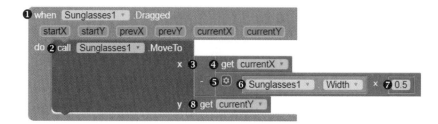

To use this method, we need to provide the coordinates for that point as the arguments for the method's x and y parameters. We'll set those coordinates using the Dragged event handler's reported event parameter arguments for the coordinates at which the user's drag of the ImageSprite ended (currentX and currentY).

To add the blocks for the x parameter, snap a - operator block ❸ to the right of the letter **x**. Then, mouse over the **currentX** event parameter and drag its **get currentX** block ❹ into the subtraction block's first operand socket.

Next, we need to subtract half the ImageSprite's width from its x-coordinate to shift it to the left, so that users can drag it from its top-center point, which is easier than dragging it from the default position, the top-left edge. To do this, drag a **x** operator block ❺ into the subtraction block's second operand socket. Then, fill the multiplication block's left operand socket by dragging in **Sunglasses1**'s **Sunglasses1.Width** block ❻ and its right operand socket with a **0.5** number block ❼. These blocks tell the app that, when a user drags Sunglasses1, it should move Sunglasses1 to the x-coordinate of the point to which the user dragged it, minus half the sprite's width.

To add the blocks for the MoveTo block's y parameter, mouse over the **currentY** event parameter and snap its **get currentY** block ❽ in next to the letter y. These blocks tell the app that, when a user drags Sunglasses1, it should move Sunglasses1 to the y-coordinate of the point to which the user dragged it.

Creating a Dragged Event Handler for All Sunglasses ImageSprites

We now need to program the exact same Dragged event response for Sunglasses2, Sunglasses3, and Sunglasses4. The following generic event handler accomplishes this task.

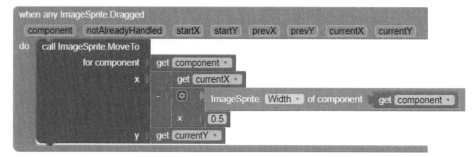

To create this any `ImageSprite Dragged` event handler, right-click the `Sunglasses1 Dragged` event handler that we just placed and select **Make Generic**. Now the blocks for step 4 move the four sunglasses `ImageSprites` around the `Canvas` when the user drags them.

Let's live-test to see these blocks work. Once you select a pair of sunglasses, they should move around the `Canvas` where you drag them. If they're not moving correctly, debug and test again. Before you test again, you may need to refresh the app in your Companion to make sure your changes take effect. You can refresh by making any change to the app in the Designer. Once this part of step 4 is working, move to the next part of the step, where we'll program the `Slider` to allow a user to change the sunglasses' sizes.

Adjusting the Width of the Selected Sunglasses

Since the 200-pixel-wide sunglasses might not fit correctly on a user's photo, we'll program the app to let the user adjust the size of the sunglasses by moving the `Slider`. The `Slider1 PositionChanged` event handler accomplishes this.

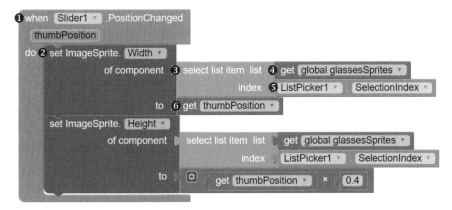

These blocks handle the action that takes place when the thumb position of the `Slider` changes from the initial value of `200` that we set in the Designer. Once the `Slider` position changes, the `whenSlider1.PositionChanged` block provides the argument for a `thumbPosition` event parameter, which holds the numerical value of the thumb position after the change.

We'll first place the blocks that set the selected `ImageSprite`'s new width to the value of `thumbPosition`, which will be somewhere between a minimum of 150 and maximum of 250, the `MinValue` and `MaxValue` properties we set for the `Slider` in the Designer. Drag `Slider1`'s `whenSlider1.PositionChanged` block ❶ to the Viewer. Next, drag a `setImageSprite.Width` block ❷ from the Any Image Sprite drawer into the event handler block. Fill the `setImageSprite.Width` block's top socket with a **select list item** block ❸, and fill that **select list item** block's top socket with a **get global glassesSprites** block ❹ and its bottom socket with the **ListPicker1.SelectionIndex** block ❺. Now, fill the `setImageSprite.Width` block's bottom socket by mousing over the **thumbPosition** event parameter and dragging in a **get thumbPosition** block ❻.

Adjusting the Height of the Selected Sunglasses

The next blocks set the height of the selected sunglasses when the Slider position changes. This is the second half of the Slider1 PositionChanged event handler.

These blocks set the pixel height of the selected sunglasses to the value of the thumbPosition multiplied by 0.4. Since the height of each pair of sunglasses is 80 pixels, which is 0.4 times the default 200-pixel width, we use this formula to maintain the same proportion between the sunglasses' height and width when the width changes. This will keep the glasses from looking distorted when a user moves the Slider.

To place these blocks, duplicate the setImageSprite.Width block. In the copy, change **Width** to **Height**, and snap the **setImageSprite.Height** block ❶ into the event handler block. Then replace the block in the **to** socket with a **×** operator block ❷. Fill that multiplication block's left operand socket by dragging in a **get thumbPosition** block ❸ and its right socket with a **0.4** number block ❹.

You've now completed step 4. Let's live-test again to see how these blocks work. You should be able to drag your selected pair of sunglasses around the Canvas and move the Slider to the left to make the glasses smaller and to the right to make them larger. If the sunglasses aren't changing size correctly or proportionally, debug and test again. Once step 4 is working, move on to step 5, where we'll program the app to let users draw on the picture.

STEP 5: LETTING USERS DRAW ON THE CANVAS

Let's now program step 5 of the app. When a user clicks the draw button, the app should display a notification that the user can draw anywhere on the Canvas. Also, when the user drags a finger on the Canvas, the app should draw on the Canvas along the finger's path, unless the user is dragging one of the sunglasses ImageSprites at the same time.

Letting Users Know They Can Draw

Here is the code that tells the user where to draw on the app screen.

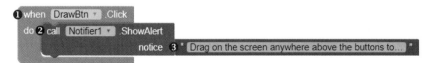

Drag the **whenDrawBtn.Click** block ❶ to the Viewer and then drag Notifier1's **callNotifier1.ShowAlert** method call block ❷ inside it.

The ShowAlert method that we're calling requires an argument for its notice parameter, which holds the information we want the Notifier to display. To provide that argument, snap in an empty string block ❸ and enter **Drag on the screen anywhere above the buttons to draw.** into its text area. Now, when the user clicks the draw button, these blocks will display an alert saying the user can drag on the Canvas to draw.

Drawing Lines on the Canvas

Next, we'll tell the app where to draw when a user drags the Canvas and in what color. The following blocks handle this Canvas1 Dragged event.

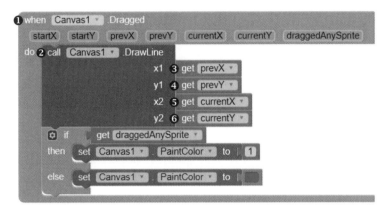

Drag the **whenCanvas1.Dragged** block ❶ to the Viewer. When the user drags the Canvas, this event handler will store the following seven event parameters as arguments: the x- and y-coordinates for the point at which the user first touched the Canvas (startX and startY), the point at which the current drag began (prevX and prevY), the point at which the current drag ended (currentX and currentY), and a Boolean value for whether the drag also dragged an ImageSprite (draggedAnySprite). We'll use a number of these values in the blocks we add to the Dragged event handler.

To add the blocks to make the app draw a line where the user dragged, drag the **callCanvas1.DrawLine** method call block ❷ into the event handler block. This method requires us to provide arguments for its x1 and y1 parameters, which hold the x- and y-coordinates for the point at which the app should start drawing, and x2 and y2 parameters, which hold the x- and y-coordinates for the point at which the app should stop drawing.

We'll get these arguments from the whenCanvas1.Dragged block. Mouse over the block's parameters and drag a **get prevX** block ❸ into the **callCanvas1 .DrawLine** block's top socket, a **get prevY** block ❹ into its second socket, a **get currentX** block ❺ into its third socket, and a **get currentY** block ❻ into its last socket. Now, when the user drags on the Canvas, these blocks direct the app to draw a line from the point at which the user begins to drag the Canvas to the point at which the user stops.

Determining the Canvas Line Color

Right now, despite the blocks we just placed, the user's lines won't show up. That's because we set the PaintColor for Canvas1 to None in the Designer. To update this, we'll have the app draw a transparent line if the user also drags an ImageSprite along with the Canvas and a red line if the user drags only the Canvas. The following if then else block directs this behavior.

Drag an **if then** block to the Viewer, click the blue mutator icon to add an **else** input to the block, and then snap the **if then else** block ❶ inside the event handler block. Now, mouse over the event handler's **draggedAny Sprite** parameter, and drag a **get draggedAnySprite** block ❷ into the **if then else** block's **if** socket, which checks whether the user has dragged one of the ImageSprites. Then, drag a **setCanvas1.PaintColor** block ❸ into the **if then else** block's **then** socket and snap a **1** number block ❹ to its right to make the color transparent. The blocks in this then socket dictate what happens if the user has also dragged an ImageSprite.

Next duplicate the **setCanvas1.PaintColor** block ❺ and drag the copy into the **if then else** block's **else** socket, which determines what happens if the user hasn't dragged an ImageSprite. Replace the **1** number block with a red color block ❻ from the Colors drawer.

Now, each time a user drags the Canvas, the app tests whether the user is also dragging a sunglasses ImageSprite. If so, the app draws an invisible line because of the 1 value we set. If the user's not dragging an ImageSprite, the app draws a red line.

Let's live-test to see how these blocks work. When you click DrawBtn, you should see an alert letting you know you can draw on the screen by dragging your finger above the buttons. When you drag on the Canvas, you should see your drawing in red on the device screen, unless you're also dragging a pair of sunglasses. If you're dragging a pair of sunglasses, you shouldn't see any lines drawn. If any part of this step isn't working correctly, debug and test again. Once step 5 is working, continue on to the next step, where we'll program the app to let users type on the picture.

STEP 6: LETTING USERS TYPE ON THE CANVAS

When a user clicks TypeBtn, the app's Notifier should open a text dialog in which the user can enter the desired text to display on the Canvas. After the user clicks OK to close the dialog, the app should display any text entered into the dialog in blue near the bottom of the Canvas.

Opening the Text Dialog for Input

The TypeBtn.Click event handler opens the text dialog to collect the entered text when the user clicks TypeBtn.

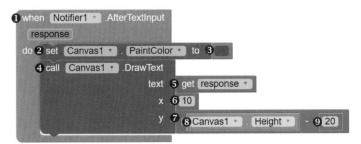

Drag the **whenTypeBtn.Click** block ❶ to the Viewer, then drag the **call Notifier1.ShowTextDialog** method call block ❷ into it. This ShowTextDialog method call requires an argument for its message parameter, which holds the message we'll show the user in the text dialog that opens; its title parameter, which holds a title that will show at the top of the text dialog; and its cancelable parameter. If the cancelable parameter were set to true, the app would add an OK and a Cancel button to the text dialog. Then, when we program the app to display the user's response later in this step, the app would display the word "Cancel" on the Canvas if the user clicked Cancel. We'll set the cancelable parameter to false instead, which adds only an OK button to the dialog. After we program the second part of this step, pressing the OK button should place the user's text on the Canvas.

For the message parameter, snap an empty string block ❸ into the message socket and enter **Type your brief message below. Click OK to close this box.** For the title parameter, snap an empty string block ❹ into the **title** socket and enter **Type on the Screen**; then, set the cancelable parameter by changing the default **true** block to **false** ❺.

Displaying the User's Input on the Canvas

Now we'll program the Notifier AfterTextInput event handler, which tells the app how and where to display the user's message (reported as the response event parameter) on the Canvas. Once the user clicks OK in the dialog, the app should set the paint color to blue and print the response at specific coordinates on the Canvas.

Drag the **whenNotifier1.AfterTextInput** block ❶ to the Viewer. Then, to set the paint color, drag the **setCanvas1.PaintColor** block ❷ into the event handler block and snap a blue color block ❸ into its right socket. After the user closes the text dialog, these blocks will change the Canvas paint color to blue.

Next, to call the method that draws the text on the Canvas, drag the **callCanvas1.DrawText** method call block into the **whenNotifier1.AfterTextInput** event handler block ❹. This `DrawText` method takes arguments for its text parameter, which is the user input it should display on the screen, and for its x and y parameters, which are the x- and y-coordinates for the point at which the input should begin to show on the `Canvas`.

To provide the argument for the method's text parameter, mouse over the **AfterTextInput** handler's **response** event parameter and drag its **get response** block ❺ into the **text** socket. For the text location, we'll set it at 10 pixels in from the `Canvas`'s left edge and 20 pixels up from the bottom so that it will display close to the bottom of the picture.

To do this, for the `callCanvas1.DrawText` block's x parameter, drag in a **10** number block ❻, and for its y parameter, drag in a - operator block ❼. Fill the subtraction block's left operand socket by dragging in the **Canvas1.Height** block ❽ and fill its right operand socket by dragging in a **20** number block ❾. This sets the x parameter to 10 and the y parameter to the height of the `Canvas` minus 20 pixels—or, in other words, 20 pixels from the bottom of the `Canvas`.

Now test these blocks. When you click `TypeBtn`, a dialog should open in which you can enter the text you want typed on the `Canvas`. Enter some text and click **OK**. You should see the text typed in blue at the left side of the `Canvas`, near the bottom. If any part of this step isn't working correctly, debug and test again. Once step 6 is working, let's program the app to allow users to save the `Canvas`.

STEP 7: LETTING USERS SAVE A PICTURE OF THE CANVAS

In this step, we'll let a user save a picture of the `Canvas`, along with any selected sunglasses, drawing, and typing, to a device as an image file. The `SaveBtn.click` event handler saves the user's creation.

First, we'll tell the app to show an alert when the user clicks the save button. Drag the **whenSaveBtn.click** block ❶ to the Viewer and then place the **callNotifier1.ShowAlert** method call block ❷ inside of it.

The next blocks provide the argument for the `ShowAlert` method's `notice` parameter, which is the notice we want the `Notifier` to show. We want that notice to tell the user where on the device the picture is saved and call the method that saves the picture while displaying the storage location.

To place these blocks, drag a **join** block ❸ into the **callNotifier1.ShowAlert** block's **notice** socket. Then fill the **join** block's top socket with an empty string block ❹ and enter **The Canvas is saved at**, including the final space, into its text area. Next, fill the **join** block's bottom socket by dragging in a **callCanvas1 .Save** method call block ❺.

Let's test these blocks. When you click SaveBtn, you should see an alert reading, "The Canvas is saved at <*file location*>." You should be able to find the file when you go to that location or look in the device's File Manager, but you may need to close the app before the file appears. Once this step works as it should, close the app on your device and move on to program the app's last step, which allows the user to clear the Canvas.

STEP 8: LETTING USERS CLEAR THE CANVAS

In this final step, we'll program the app to let the user clear the sunglasses, as well as all drawings and typed text, from the Canvas, so that only the background image remains. The TrashBtn.click event handler clears the Canvas.

These blocks call the clearSprite procedure, which hides all ImageSprites, like it did when we used it in the AfterPicker event handler. Then they call the built-in Canvas Clear method, which clears all text and drawings from the Canvas, when the user clicks the trash button.

Drag the whenTrashBtn.click block ❶ to the Viewer and then drag the call clearSprite block ❷ into it. Next drag a callCanvas1.Clear block ❸ into the event handler block under the call clearSprite block.

Now let's test the completed app! Open the app on your device. You should see the TakePicBtn. Click it, and your camera should open. When you take a picture, that picture should become the background of the top portion of your app screen. Below the picture, you should see the ListPicker, with the Slider and the draw, type, save, and trash buttons beneath it.

When you click Choose Glasses, you should see all four sunglasses options listed, and, when you select a pair, the ListPicker should close and your selected sunglasses should appear at the top of the Canvas. You should be able to drag the glasses around the Canvas and use the Slider to change their size.

Test each button. When you click DrawBtn, the app should display the notice telling you that you can draw on the Canvas, and when you drag your finger along the Canvas at any place except on the ImageSprite, a red line should appear. When you click TypeBtn, a dialog should open prompting you to type a message, and after you type it and click OK, your message should appear in blue at the bottom of the Canvas.

When you click SaveBtn, you should see a notice alerting you that a picture of the Canvas has been saved at a given location on your device. When you click TrashBtn, the ImageSprite and all text and drawing should disappear from the Canvas. If you placed your blocks correctly, you've successfully created the "Virtual Shades" app!

SUMMARY

In this chapter, you built the "Virtual Shades" app, which lets users try on sunglasses, draw and type on the Canvas, and save a picture of the Canvas to their devices to keep and share.

Along the way, you learned how to set the Canvas line size and colors to prepare for drawing, let users take a selfie through the app and make it the app background, let users adjust components with the Slider, and use the Notifier component to not only display information but also collect user input.

You also practiced declaring and initializing list variables, creating procedures, providing required arguments for built-in methods with parameters, using comparison operators to test conditions, and adding conditionals and loops to control app flow.

ON YOUR OWN

Modify and extend the "Virtual Shades" app by working on the following exercises. Be sure to adjust your algorithm to plan the changes before you add any components or blocks, and save the new versions of the apps you create. You can find solutions online at *https://nostarch.com/ programwithappinventor/*.

1. Change the app so that it allows users to choose the Canvas paint color to draw and type on the Canvas.
2. Change the app so that it allows users to speak the words they want entered on the Canvas.
3. Extend the app so that users can opt not to take a background photo and instead create art on a blank Canvas.

APP INVENTOR COMPONENTS AND BUILT-IN BLOCKS

 App Inventor includes drawers of components you can add to your app as well as built-in blocks you'll use to set its general behaviors. This appendix gives a current overview of both. Note that the App Inventor developers add new components and update blocks from time to time, and they mention those changes in the Welcome splash screen that you see when you log in.

APP INVENTOR COMPONENTS

The App Inventor Designer window includes a *Palette* pane containing drawers of components you can add to your app (see Figure A-1).

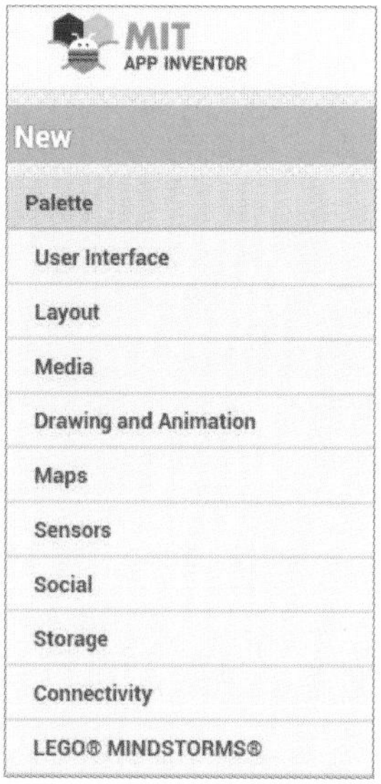

Figure A-1: The App Inventor
Designer window's Palette pane
showing types of components

Components are all the visible and non-visible elements you can use in your app, including those that set its look and feel and add exciting functionality. All components have properties you can set and/or actions you can program.

USER INTERFACE COMPONENTS

The User Interface components, shown in Figure A-2, control the elements users see and touch in your app.

These components help users understand what you want them to do and then accomplish those tasks. Several User Interface components also allow you to collect the user input necessary for your app to work.

Button Displays a button or picture on the screen that can detect the user's click. You often will add text to a Button to let users know what will happen when they click it.

CheckBox Lets users select one or more options presented.

DatePicker Displays a button for the user to click and select a date. Having users input dates with the DatePicker ensures the dates are formatted consistently when the app receives them, which is easier for the app to process.

Image Shows an image on the screen.

Label Displays text on the screen.

ListPicker Shows a button for the user to click to see a list of text items to choose from. You can make the list of items searchable.

ListView Displays a list of text items to the user.

Notifier Displays a temporary alert or a message that prompts the user to confirm, choose, enter text, or wait before continuing to interact with the app. The Notifier component is also used to log errors or messages.

PasswordTextBox Prompts users to enter text and other characters that shouldn't be displayed as they're being typed, like a password. It's the same as the TextBox component, except each character the user enters into the box displays as an asterisk (*).

Slider Displays a progress bar with a "thumb" the user can drag left or right to set its position. A Slider allows the user to quickly select a value from a range.

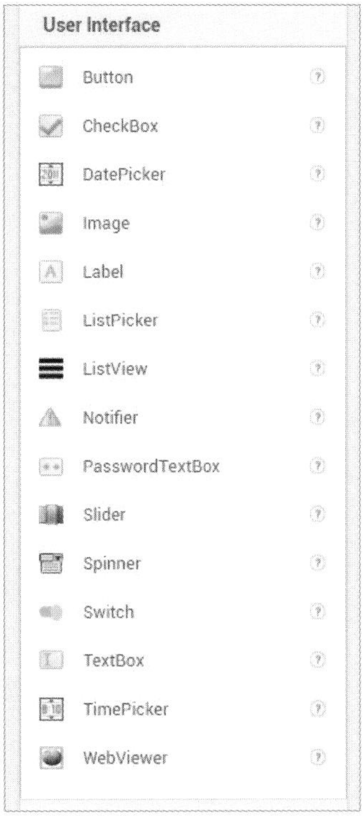

Spinner Shows a new screen containing a drop-down list of text items.

Switch Displays a toggle switch on the screen for the user to click.

TextBox Displays a box where the user can enter text, numbers, and other characters. The characters the user types into the box will be visible.

TimePicker Displays a button for the user to click and select a time. Having users input times with the TimePicker ensures the times are formatted consistently when the app receives them, which is easier for the app to process.

WebViewer Shows a web page within the app.

Figure A-2: The User Interface components drawer

LAYOUT COMPONENTS

The formatting components available in the Layout drawer, shown in Figure A-3, help you arrange visible components (such as those from the User Interface drawer) in an orderly and visually appealing way.

HorizontalArrangement   Displays components from left to right.

HorizontalScrollArrangement   Displays components side-by-side in a container that scrolls from left to right.

TableArrangement   Displays components as if they are laid out in a table (that is, in rows and columns).

VerticalArrangement   Displays components stacked, one below another.

VerticalScrollArrangement   Displays components stacked, one below another, in a container that scrolls from top to bottom.

Figure A-3: The Layout components drawer

MEDIA COMPONENTS

You'll find the Camera and other components that access a device's media functions in the Media drawer, shown in Figure A-4.

Camcorder   Lets the app record a video using the device's camcorder.

Camera   Allows the app to take a picture using the device's camera.

ImagePicker   Displays a button the user clicks to choose images from the device's image gallery.

Player   Plays longer audio files, such as songs, and controls phone vibration.

Sound   Plays short sound files, such as sound effects, and also vibrates.

SoundRecorder   Lets your app record audio.

SpeechRecognizer   Converts speech to text that can be used in an app.

TextToSpeech   Lets the app read text aloud.

VideoPlayer   Plays videos.

YandexTranslate   Enables the app to translate words and sentences between different languages using the Yandex.Translate service.

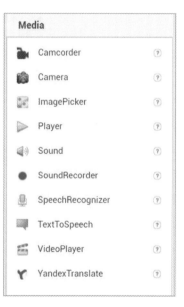

Figure A-4: The Media components drawer

DRAWING AND ANIMATION COMPONENTS

The Drawing and Animation components, shown in Figure A-5, enable users to draw on the screen and let you include moving images in your app.

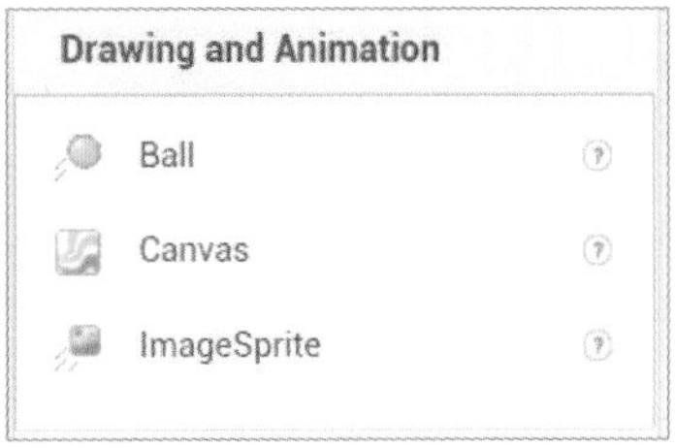

Ball   A round *sprite* (a small two-dimensional image) that, when placed on a Canvas, can move, react to touches and drags, and interact with other sprites and with the edge of the Canvas.

Canvas   A two-dimensional, touch-sensitive rectangular panel that allows you to draw on the screen or place or move sprites around on the screen.

ImageSprite   A sprite that, when placed on a Canvas, can move, react to touches and drags, and interact with other sprites and with the edge of the Canvas.

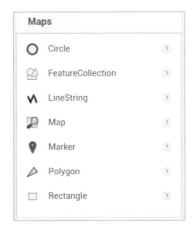

Figure A-5: The Drawing and Animation components drawer

MAPS COMPONENTS

To create apps that focus on points on a map, use the tools in the Maps components drawer, shown in Figure A-6.

Circle   Draws a circle of a given radius around a point on a Map.

FeatureCollection   Groups one or more Maps components together.

LineString   Draws a sequence of line segments on a Map.

Map   A two-dimensional container that displays a map as its background. You must first add this component to place any other Maps component on the screen.

Marker   Places an icon at a point to indicate information on a Map.

Polygon   Draws an arbitrary shape on a Map.

Rectangle   Draws a rectangle on a Map bounded by north, south, east, and west edges.

Figure A-6: The Maps components drawer

SENSORS COMPONENTS

The Sensors components drawer, shown in Figure A-7, offers several fun and powerful components like the Clock component, which enables the timed automatic action necessary for many games to run, and the LocationSensor, which checks an object's location from time to time. Note that all Sensors are *non-visible components*, meaning when you add them to your app, your users will not see them on the screen.

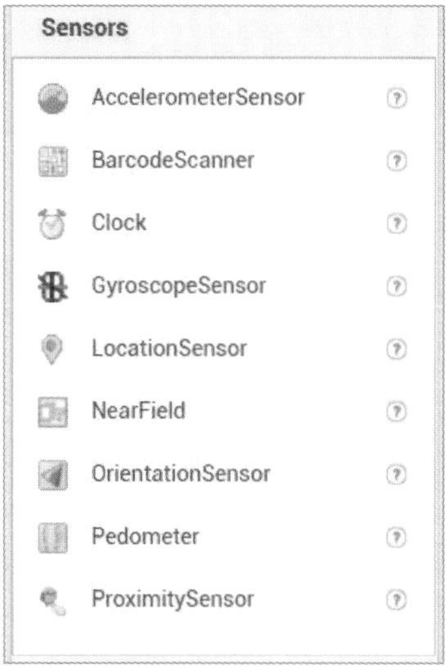

Figure A-7: The Sensors components drawer

AccelerometerSensor   Detects a device's shaking and measures its acceleration in three dimensions.

BarcodeScanner   Reads barcodes, if the device has a barcode scanner app installed.

Clock   Uses the device's internal clock to fire a timer at regularly set intervals and perform time calculations, manipulations, and conversions.

GyroscopeSensor   Measures angular velocity in three dimensions.

LocationSensor   Provides location information, including longitude, latitude, altitude (if supported by the device), speed (if supported by the device), and address. It also can *geocode*, or convert a given address to latitude and longitude.

NearField   Provides *near-field communication* capability with other devices (if supported by the device).

OrientationSensor   Provides information about the device's physical orientation in three dimensions.

Pedometer   Attempts to determine if a step has been taken and estimate the distance traveled.

ProximitySensor   Measures the distance of an object to the device's view screen.

SOCIAL COMPONENTS

Components in the Social drawer, shown in Figure A-8, enable your users to get social by communicating with others.

ContactPicker Displays a button the user clicks to choose a contact stored in the device.

EmailPicker Includes a searchable text box where a user can begin to enter a contact's name or email address, prompting the device to show a drop-down menu of contacts for the user to choose from.

PhoneCall Enables the app to make a phone call to a specified number.

PhoneNumberPicker Displays a button the user clicks to choose a contact phone number stored in the device.

Sharing Lets the user share files and/or messages between the app and other apps installed on the device, such as email and message apps.

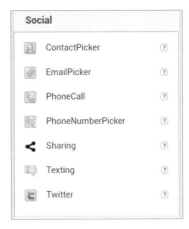

Figure A-8: The Social components drawer

Texting Enables the app to send a specified text message to a specified phone number.

Twitter Allows users to communicate with Twitter to send and search tweets and more.

STORAGE COMPONENTS

In the Storage drawer, shown in Figure A-9, you'll find components that enable your app to remember information even after it closes.

File Enables the app to store and retrieve files.

FusiontablesControl Lets the app communicate with Google Fusion Tables to share, query, and store data into tables.

TinyDB Enables the app to save and retrieve information stored directly on the device.

TinyWebDB Allows the app to communicate with a web service to store and retrieve information that can be shared with other devices.

Figure A-9: The Storage components drawer

CONNECTIVITY COMPONENTS

The Connectivity components, shown in Figure A-10, enable your app to connect to other apps, other devices, and the internet.

ActivityStarter Enables the app to launch an activity, such as another app or a web browser.

BluetoothClient Allows the app to connect to another Bluetooth-connected device.

BluetoothServer Allows the app to accept a connection from another Bluetooth-connected device.

Web Enables the app to access data from a web service and use it in the app.

Connectivity		
⚡	ActivityStarter	⑦
◼	BluetoothClient	⑦
◼	BluetoothServer	⑦
⚫	Web	⑦

Figure A-10: The Connectivity components drawer

OTHER COMPONENTS

Beneath the Connectivity drawer, you'll find components you can use to program LEGO® MINDSTORMS® robots, as well as experimental and extension components that evolve and change. Be sure to pay attention to the Welcome splash screen that opens when you log in to App Inventor, which displays information about notable updates.

APP INVENTOR'S BUILT-IN BLOCKS

The App Inventor Blocks Editor window includes a Blocks pane containing drawers of built-in blocks that add general behaviors to your app (see Figure A-11).

Control blocks Control an app's flow of action. Some allow an app to compare values and test conditions (the if then and if then else blocks) and repeat actions (the for each number from to, for each item in list, and while test blocks). Other blocks open and close and pass information between screens and transfer text to and from other apps on the device.

Logic blocks Two types of Logic blocks (true and false) are used to set the values of Boolean properties and variables, which have only two possible values: true and false. The other blocks are logical or Boolean operators (the not, =, ≠, and, and or blocks) used to create Boolean

Figure A-11: Built-in block categories

expressions—expressions that evaluate as either true or false—to compare values and test conditions. These operators are often used with Control blocks to control an app's flow of action.

Math blocks   Some blocks assign regular and random numerical values. Some are arithmetic operators that perform arithmetic, algebra, and trigonometry calculations. Other blocks are relational or comparison operators (the =, ≠, >, ≥, <, and ≤ blocks) that compare values. These operators also are often used with Control blocks to control an app's flow of action.

Text blocks   Add, join, count characters, compare, change, and search within strings. Strings can contain letters, numbers, spaces, and special characters.

Lists blocks   Create, count items, change, search, select from, copy, search within, and otherwise handle lists of items.

Colors blocks   Add color to components and also make and split colors.

Variables blocks   Initialize global and local variables and get and set values for the variables you create.

Procedures blocks   Create and call procedures.

INDEX